Modeling Hydrodynamics, Water Temperature, and Water Quality in the Klamath River Upstream of Keno Dam, Oregon, 2006–09

Annett B. Sullivan and Stewart A. Rounds, U.S. Geological Survey; Michael L. Deas, Watercourse Engineering, Inc.; Jessica R. Asbill, Bureau of Reclamation; Roy E. Wellman, Marc A. Stewart, and Matthew W. Johnston, U.S. Geological Survey; and I. Ertugrul Sogutlugil, Watercourse Engineering, Inc.

Prepared in cooperation with the Bureau of Reclamation

Scientific Investigations Report 2011-5105

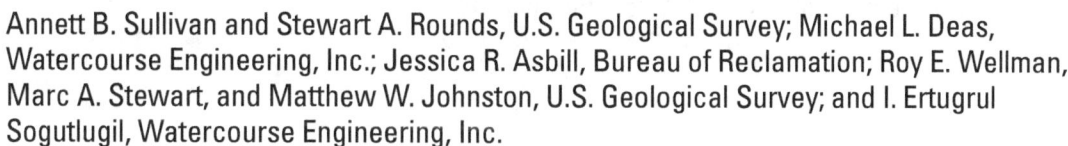

U.S. Department of the Interior
U.S. Geological Survey

U.S. Department of the Interior
KEN SALAZAR, Secretary

U.S. Geological Survey
Marcia K. McNutt, Director

U.S. Geological Survey, Reston, Virginia: 2011

For more information on the USGS—the Federal source for science about the Earth, its natural and living resources, natural hazards, and the environment, visit http://www.usgs.gov or call 1–888–ASK–USGS.

For an overview of USGS information products, including maps, imagery, and publications, visit http://www.usgs.gov/pubprod

To order this and other USGS information products, visit http://store.usgs.gov

Suggested citation:
Sullivan, A.B., Rounds, S.A., Deas, M.L., Asbill, J.R., Wellman, R.E., Stewart, M.A., Johnston, M.W., and Sogutlugil, I.E., 2011, Modeling hydrodynamics, water temperature, and water quality in the Klamath River upstream of Keno Dam, Oregon, 2006–09: U.S. Geological Survey Scientific Investigations Report 2011-5105, 70 p.

Contents

Contents–Continued

Figures

Figures—Continued

Tables

Conversion Factors, Datums, and Abbreviations and Acronyms

Conversion Factors

Inch/Pound to SI

Multiply	By	To obtain
foot (ft)	0.3048	meter (m)
mile (mi)	1.609	kilometer (km)
acre-foot (acre-ft)	1,233	cubic meter (m^3)
cubic foot per second (ft^3/s)	0.02832	cubic meter per second (m^3/s)
pound (lb)	0.45359	kilogram (kg)

SI to Inch/Pound

Multiply	By	To obtain
micrometer (µm)	0.00003937	inch (in.)
meter (m)	3.281	foot (ft)
square meter (m^2)	1,550.0	square inch (in^2)
liter (L)	33.82	ounce, fluid (fl. oz)
liter (L)	0.2642	gallon (gal)
cubic meter (m^3)	264.2	gallon (gal)
liter (L)	61.02	cubic inch (in^3)
cubic meter (m^3)	35.31	cubic foot (ft^3)
Watt (W)	3.41	British thermal unit per hour (BTU/h)

Temperature in degrees Celsius (°C) may be converted to degrees Fahrenheit (°F) as follows:

$$°F = (1.8 \times °C) + 32.$$

Specific conductance is given in microsiemens per centimeter at 25 degrees Celsius (µS/cm at 25 °C).

Concentrations of chemical constituents in water are given either in milligrams per liter (mg/L), which is approximately equivalent to parts per million (ppm), or micrograms per liter (µg/L), which is approximately equivalent to parts per billion (ppb).

Datums

Elevation refers to distance above the vertical datum. A local vertical datum (UKLVD) is used, established by the Bureau of Reclamation. For the purpose of this report, the conversion is UKLVD – 1.78 ft = NGVD29.

Horizontal coordinate information is referenced to the North American Datum of 1983 (NAD 83).

Conversion Factors, Datums, and Abbreviations and Acronyms—Continued

Abbreviations and Acronyms

AFA	*Aphanizomenon flos aquae*
BOD	biochemical oxygen demand
BOD5	oxygen consumed over 5 days as a result of BOD
DOM	dissolved organic matter
GIS	geographic information system
>	greater than
ISS	inorganic suspended sediment
KR	Klamath River
LDOM	labile dissolved organic matter
LPOM	labile particulate organic matter
<	less than
NPDES	National Pollutant Discharge Elimination System
NH4	ammonia
NO3	nitrate
NTU	nephelometric turbidity units
NWIS	National Water Information System
ODEQ	Oregon Department of Environmental Quality
PDT	Pacific Daylight Time
POM	particulate organic matter
PO4	orthophosphorus
PST	Pacific Standard Time
RDOM	refractory dissolved organic matter
RPOM	refractory particulate organic matter
SC	specific conductance
SOD	sediment oxygen demand
T	water temperature
TMDL	total maximum daily load
TDS	total dissolved solids
TSS	total suspended solids
UKL	Upper Klamath Lake
USGS	U.S. Geological Survey
VSS	volatile suspended solids
WWTP	wastewater treatment plant

Modeling Hydrodynamics, Water Temperature, and Water Quality in the Klamath River Upstream of Keno Dam, Oregon, 2006–09

By Annett B. Sullivan and Stewart A. Rounds, U.S. Geological Survey; Michael L. Deas, Watercourse Engineering, Inc.; Jessica R. Asbill, Bureau of Reclamation; Roy E. Wellman, Marc A. Stewart, and Matthew W. Johnston, U.S. Geological Survey; and I. Ertugrul Sogutlugil, Watercourse Engineering, Inc.

Executive Summary

A hydrodynamic, water temperature, and water-quality model was constructed for a 20-mile reach of the Klamath River downstream of Upper Klamath Lake, from Link River to Keno Dam, for calendar years 2006–09. The two-dimensional, laterally averaged model CE-QUAL-W2 was used to simulate water velocity, ice cover, water temperature, specific conductance, dissolved and suspended solids, dissolved oxygen, total nitrogen, ammonia, nitrate, total phosphorus, orthophosphate, dissolved and particulate organic matter, and three algal groups. The Link–Keno model successfully simulated the most important spatial and temporal patterns in the measured data for this 4-year time period. The model calibration process provided critical insights into water-quality processes and the nature of those inputs and processes that drive water quality in this reach. The model was used not only to reproduce and better understand water-quality conditions that occurred in 2006–09, but also to test several load-reduction scenarios that have implications for future water-resources management in the river basin.

The model construction and calibration process provided results concerning water quality and transport in the Link–Keno reach of the Klamath River, ranging from interesting circulation patterns in the Lake Ewauna area to the nature and importance of organic matter and algae. These insights and results include:

- Modeled segment-average water velocities ranged from near 0.0 to 3.0 ft/s in 2006 through 2009. Travel time through the model reach was about 4 days at 2,000 ft³/s and 12 days at 700 ft³/s flow. Flow direction was aligned with the upstream–downstream channel axis for most of the Link–Keno reach, except for Lake Ewauna. Wind effects were pronounced at Lake Ewauna during low-flow conditions, often with circulation in the form of a gyre that rotated in a clockwise direction when winds were towards the southeast and in a counterclockwise direction when winds were towards the northwest.

- Water temperatures ranged from near freezing in winter to near 30 °C at some locations and periods in summer; seasonal water temperature patterns were similar at the inflow and outflow. Although vertical temperature stratification was not present at most times and locations, weak stratification could persist for periods up to 1–2 weeks, especially in the downstream parts of the reach. Thermal stratification was important in controlling vertical variations in water quality.

- The specific conductance, and thus density, of tributaries within the reach usually was higher than that of the river itself, so that inflows tended to sink below the river surface. This was especially notable for inflows from the Klamath Straits Drain, which tended to sink to the bottom of the Klamath River at its confluence and not mix vertically for several miles downstream.

- The model was able to capture most of the seasonal changes in the algal population by modeling that population with three algal groups: blue-green algae, diatoms, and other algae. The blooms of blue-green algae, consisting mostly of *Aphanizomenon flos aquae* that entered from Upper Klamath Lake, were dominant, dwarfing the populations of the other two algae groups in summer. A large part of the blue-green algae population that entered this reach from upstream tended to settle out, die, and decompose, especially in the upper part of the Link–Keno reach. Diatoms reached a maximum in spring and other algae in midsummer.

- Organic matter, occurring in both dissolved and particulate forms, was critical to the water quality of this reach of the Klamath River, and was strongly tied to nutrient and dissolved-oxygen dynamics. Dissolved and particulate organic matter were subdivided into labile (quickly decaying) and refractory (slowing decaying) groups for modeling purposes. The particulate matter in summer, consisting largely of dead blue-green algae, decayed quickly. Consequently, this particulate matter exerted a high oxygen demand over short periods and contributed strongly to low dissolved-oxygen conditions present during summer and fall. Particulate matter in winter and dissolved organic matter throughout the year was largely refractory (slow to decay). The slower decay rate of this refractory material translates to less oxygen demand over short periods, but also will manifest itself as higher oxygen demand downstream of Keno Dam.

- The decay and settling of algae and particulate organic matter in the upper part of the Link–Keno reach of the Klamath River has important implications for nutrients. Decay releases nitrogen and phosphorus from particulate forms into dissolved forms such as ammonia, which had elevated concentrations in the downstream part of this reach in summer. Dissolved nutrients showed consistent seasonal patterns that were simulated well by the model. Ammonia concentrations were highest in midsummer and winter and lowest in spring. Nitrate concentrations were highest in winter and lowest in summer. Orthophosphorus concentrations were at their maximum in midsummer and lowest in winter. Comparing modeled hourly nutrient loads at the Link River inflow and the Keno Dam outflow, the Link–Keno reach and its tributaries were a source of total nitrogen and total phosphorus to downstream reaches in early spring and a sink in summer.

- Dissolved-oxygen concentrations were near saturation in winter, but periods of supersaturation could occur in spring and early summer as oxygen was produced by algal photosynthesis. In mid- to late summer, oxygen sources were overwhelmed by oxygen sinks, especially the decay of organic matter in the water column and river bottom. Extensive anoxia occurred during this period. The sediment oxygen demand was dynamic and represented a relatively fast decomposition of materials deposited during that same year. The labile material was eventually exhausted and reaeration from the atmosphere allowed the system to slowly return towards oxygen saturation in fall. The model simulated the general temporal and spatial patterns in dissolved oxygen, although the inclusion of macrophytes and additional information on reaeration processes, organic matter, and algal dynamics could improve the simulation of dissolved oxygen.

- Calendar years 2007 and 2008 had more extensive datasets than 2006 and 2009. The models built with less extensive input data were still able to reproduce the patterns in the measured data reasonably well. These findings underline the importance of using results from the 2007 and 2008 detailed field data and

experimental work to determine robust model rates, stoichiometry relations, and other parameters that can be applied successfully to years with less data and with different conditions.

- The 2006–09 models were applied to examine the effects of several reduced-loading scenarios consistent with total maximum daily load (TMDL) targets. The water quality of the Link River inflow was modified in one scenario so that it met the in-lake phosphorus targets of the Upper Klamath Lake TMDL. Point and nonpoint sources along the Klamath River were set to be in compliance with their Klamath River TMDL allocations in another scenario. Results from those scenarios indicated that dissolved-oxygen conditions improved the most when Link River loads were reduced; depending on year, average June through October dissolved-oxygen concentrations increased between 1.9 and 3.2 mg/L. Similarly, ammonia concentrations improved the most under this scenario, with an average June through October concentration decrease between 0.20 and 0.34 mg/L. Orthophosphorus concentrations were decreased significantly in both scenarios that reduced concentrations from Link River and scenarios that reduced concentrations from in-reach point and nonpoint sources, with June through October concentration decreases between 0.02 and 0.06 mg/L.

The calibrated models are useful tools that reproduce the most important water-quality processes occurring in the Link–Keno reach of the Klamath River. These models are accurate enough to provide insights into the nature of those processes and the probable effects of proposed management and water-quality improvement strategies.

Introduction

Background

The Klamath River flows about 255 mi (410 km) from the outlet of Upper Klamath Lake through southern Oregon and northern California to the Pacific Ocean. The first 1-mi reach, just downstream of Upper Klamath Lake, is named Link River (fig. 1). The Klamath River proper begins at the mouth of Link River, and river stage in the next 20 mi is controlled by Keno Dam. Water quality in the Link River to Keno Dam reach of the Klamath River has been classified as "very poor" by the State of Oregon (Mrazik, 2007) and was designated as "water quality limited" on the State of Oregon's 303(d) list for exceeding ammonia and dissolved-oxygen criteria throughout the year, and pH and chlorophyll *a* criteria in summer (Oregon Department of Environmental Quality, 2007). Although no numeric temperature criteria has been set for this reach, allocations have been established for the various

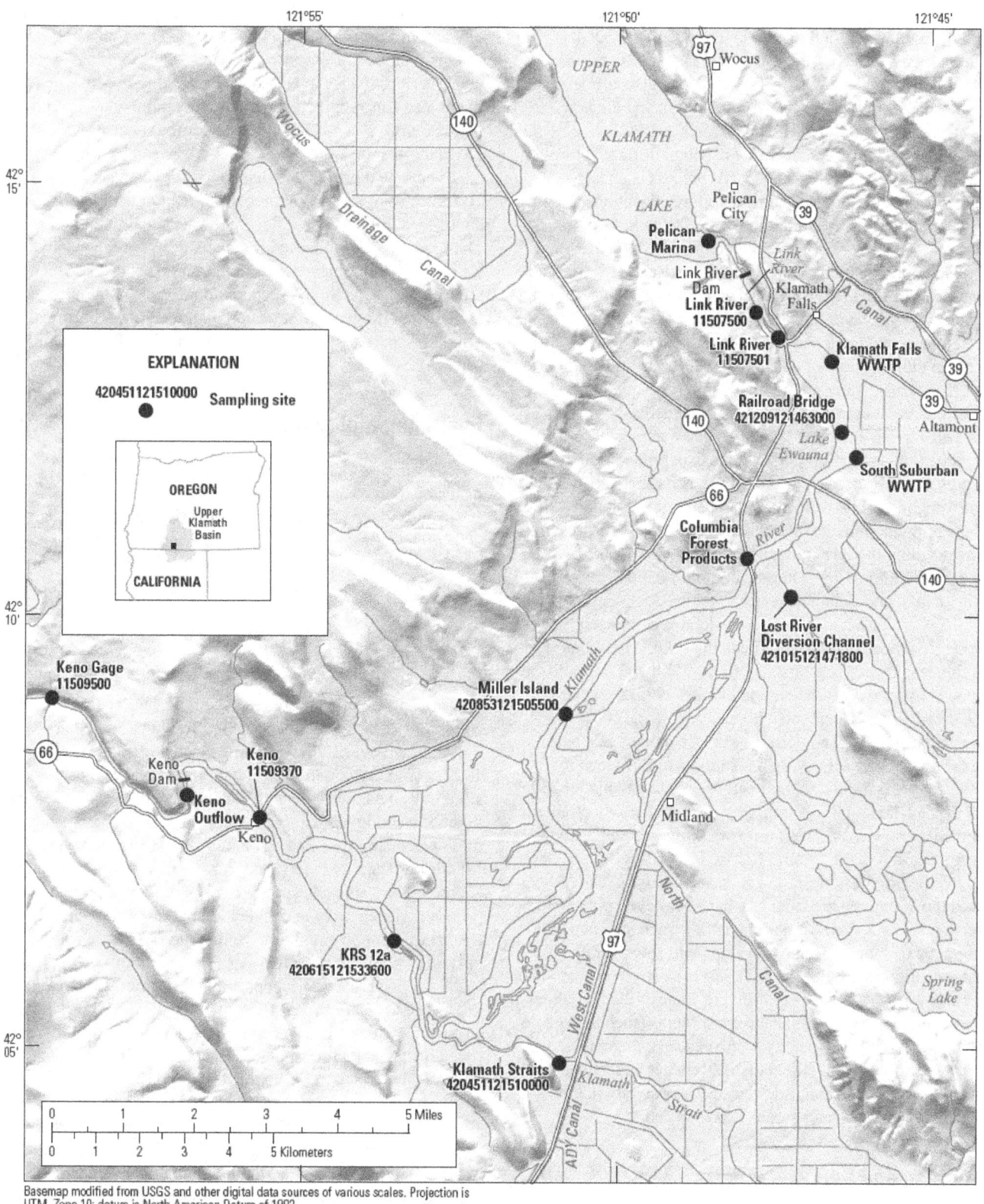

Figure 1. Location of the study reach and sampling locations in the Klamath River upstream of Keno Dam, Oregon. WWTP, wastewater treatment plant.

inflows to the reach to address the in-reach cool water criterion and the cold water total maximum daily load (TMDL) criterion downstream of Keno Dam. Fish die-offs in summer are not uncommon (W. Tinniswood, Oregon Department of Fish and Wildlife, written commun., 2006), and are noted to occur during conditions with poor water quality.

Concerns for aquatic life are driving efforts to improve water quality in this reach. A TMDL process is underway that will specify nutrient and temperature allocations to point and nonpoint sources along this reach. The U.S. Geological Survey Oregon Water Science Center (USGS), the Bureau of Reclamation Klamath Basin Area Office (Reclamation), and Watercourse Engineering, Inc. began a study in 2006 to better understand the water quality and the processes controlling water quality in the Klamath River upstream of Keno Dam, with an ultimate goal of constructing accurate and predictive hydrodynamic, water temperature, and water quality models for the 2006–09 period. The field dataset collected for this study (Sullivan and others, 2008, 2009) was critical for model construction and calibration, and the experimental studies (Poulson and Sullivan, 2010; Sullivan and others, 2010; Deas and Vaughn, 2011) contributed to a better understanding of some of the dominant water quality processes in this reach. The resulting models allow the effectiveness of various options for improving water quality to be evaluated in a quantitative and reliable manner.

Several previous efforts to model this complex river reach have been completed. The first CE-QUAL-W2 models of this reach were developed by CH2M-Hill and Portland State University (CH2M-Hill and Wells, 1995) for years 1990 and 1992. Watercourse Engineering, Inc. (2003; PacifiCorp, 2005) used portions of those models as a basis for constructing a CE-QUAL-W2 model of this reach for calendar years 2000–2004, as part of an effort to model 250 mi of the Klamath River from Link Dam to near its discharge to the Pacific Ocean. That model was constructed to support an application to the Federal Energy Regulatory Commission by PacifiCorp for hydropower relicensing. Tetra Tech (2009) used the Watercourse model as a basis for developing a version of the Klamath River model for calendar years 2000 and 2002 to support TMDL development. However, the datasets used to develop these earlier Link River to Keno Dam models were not sufficient to capture some of the fundamental characteristics of processes and constituents (such as organic matter) driving water quality in this reach (for example, Rounds and Sullivan, 2009, 2010). To better understand and predict some of the major drivers of water quality in this reach, new datasets and models were needed.

Purpose and Scope

The purpose of this study was to develop a model of the Klamath River from Link River to Keno Dam that could (1) simulate stage, flow, velocity, temperature, and water quality, (2) provide information on processes that control water

quality, (3) provide insight that would aid in the development of water quality monitoring plans, and (4) predict changes in velocity, temperature, and water quality that are likely to occur under various management and water quality improvement scenarios. This report addresses the first three goals and begins on the fourth; plans for the simulation of a range of additional scenarios is underway.

Separate models were developed for calendar years 2006, 2007, 2008, and 2009. All models were built using version 3.6 of the CE-QUAL-W2 flow and water-quality model. The models were calibrated for stage, flow, water velocity, ice cover, water temperature, inorganic suspended sediment, three algal groups (blue-green, diatom, other), total nitrogen, particulate nitrogen, nitrate, ammonia, total phosphorus, orthophosphorus, particulate carbon, and dissolved organic carbon.

Environmental Setting

The upper Klamath River lies on a broad volcanic plateau between the Cascade Range to the west and the Basin and Range province to the east (Gannett and others, 2007). The climate is semiarid, with dry summers; most precipitation occurs in fall and winter. Upper Klamath Lake is just upstream of the study reach (fig. 1); it is a large 89.6 mi^2 (232 km^2), shallow [full pool mean depth 9 ft (2.8 m)] lake that tends to have dense summer blooms of the blue-green alga *Aphanizomenon flos-aquae* (AFA). Studies of sediment cores taken from the lake indicate that the lake historically was eutrophic, with a more recent shift to hypereutrophic conditions (Eilers and others, 2003; Bradbury and others, 2004). Link River Dam was constructed at the lake outlet in 1921, and in combination with partial removal of a bedrock sill, allowed a larger range of lake storage to be controlled for hydropower and irrigation diversion, as well as management of downstream flows in the Klamath River. In recent years, Link River Dam has been used to maintain the stage in Upper Klamath Lake at levels specified by Biological Opinions to protect two species of endangered fish: the Lost River sucker (*Deltistes luxatus*) and the shortnose sucker (*Chasmistes brevirostris*).

Link River flows for approximately 1 mi downstream of Link River Dam to the start of the Klamath River at Lake Ewauna, a wide and shallow area near Klamath Falls, Oregon (fig. 1). Channel widths in Lake Ewauna can approach 2,500 ft (about 800 m); in the rest of the reach down to Keno Dam, channel widths are 300–1,000 ft (about 100–300 m). Channel depths in the Link River to Keno Dam reach range up to approximately 20 ft (6 m). Studies of core samples have determined that organic matter, including wood chips from historical wood processing operations, is common on the river bottom, especially at the upper end of the reach. Locally, bedrock is close to the river bottom, especially near Keno (Eilers and Raymond, 2003; Raymond and Eilers, 2004; Eilers and Raymond, 2005).

The major inflow to the reach is at Link River (table 1). Other inflows in 2006–09 include Klamath Straits Drain and three point sources with National Pollutant Discharge Elimination System (NPDES) permits. These include two wastewater treatment plants (WWTP; Klamath Falls and South Suburban) and Columbia Forest Products (fig. 1). Gaged outflows include withdrawals through the Ady and North Canals to supply water for irrigation and wildlife refuges, and flow through Keno Dam to downstream reaches of the Klamath River. The Lost River Diversion Channel conveys water between the Klamath and Lost River systems. Water can be conveyed through the Lost River Diversion Channel in either direction, but typical operations include diversion to the Klamath River in the winter and diversion from the Klamath River in the summer.

The blue-green algae AFA was the most common algae in this part of the upper Klamath River from early summer into fall; the source of most AFA was from Upper Klamath Lake (Sullivan and others, 2008, 2009). Diatoms were at their maximum during spring and other algae, including cryptophytes and green algae, were present during various times of the year. The most frequently identified zooplankton in 2007–08 were cladocerans, copepods, and rotifers, especially *Daphnia pulicaria*, *Chydorus sphaericus*, copepod nauplii, cyclopoid copepodites, *Keratella hiemalis*, and *Euchlanis dilatata*. Fish species in the reach during a 2002–03 survey include fathead minnow (*Pimephales promelas),* blue chub (*Gila coerulea*), tui chub (*Gila bicolor*), shortnose sucker, and Lost River sucker among others (Terwilliger and others, 2004). Although anadromous fish are no longer present, this reach has historically supported such fish runs, and the reintroduction of Chinook salmon (*Onchorhynchus tshawytscha*) is being studied (Dunsmoor and Huntington, 2006; Oregon Department of Fish and Wildlife, 2008).

Keno Dam, approximately 20 mi downstream of Link River, is a stage- and flow-regulating facility constructed in 1967 (replacing a previous regulation facility) at the location of a natural basalt structure. Normal full pool elevation at Keno Dam is 4,085 ft (1,245 m) above sea level, and the total storage capacity of this reach is reported as 18,500 acre-ft (22.8 million m^3) (PacifiCorp, 2002). Four other dams downstream on the Klamath River are being considered for removal, but Link River Dam and Keno Dam are slated to remain to regulate stage and flow.

Methods

Model Description

The Link–Keno model was constructed with CE-QUAL-W2 version 3.6 (Cole and Wells, 2008), a two-dimensional, laterally averaged hydrodynamic and water-quality model from the U.S. Army Corps of Engineers and Portland State University. CE-QUAL-W2 can simulate water level, flow, water velocity, water temperature, ice cover, and many water-quality constituents, including total dissolved solids, dissolved oxygen, pH, nutrients, particulate and dissolved organic matter, and algae. It has been applied successfully to hundreds of rivers, lakes, and reservoirs around the world. At a reach scale, a long, narrow, pooled river is typically a good candidate for a two-dimensional, laterally averaged model. Longitudinal and vertical variations are clearly exhibited in field data and indicate that simulation of longitudinal and vertical differences are useful. Although there are certain areas of the river that exhibit lateral (bank-to-bank) variability, these conditions are generally local in character (Vaughn and Deas, 2006; Sullivan and others, 2009). This does not preclude the importance of lateral variability, but on a reach scale (versus a local scale), the CE-QUAL-W2 model was deemed appropriate.

The Link–Keno model was developed in several steps. First, a model grid was constructed to represent the course and morphology of the reach. Subsequently, necessary data were collected and formatted to provide meteorological, hydrological, and water temperature and water-quality boundary conditions. Prior to model calibration, representative values for many parameters (for example, rate constants, coefficients, stoichiometric ratios, and so forth) were defined through data derivations, results from experimental work, or from the literature.

The water budget for the Link–Keno reach was calibrated by comparing measured and modeled water levels to estimate the gains or losses of water from ungaged inflows, outflows, and groundwater interactions. Water temperature and water-quality constituents were calibrated by

Table 1. Mean measured inflows and outflows in the Klamath River above Keno Dam, Oregon, calendar years 2006–09.

[**Abbreviations:** WWTP, wastewater treatment plant; ft^3/s, cubic feet per second]

Name	2006	2007	2008	2009
Inflows (ft^3/s)				
Link River	1,929	1,165	1,075	997
Klamath Falls WWTP	5.0	3.5	3.2	2.8
South Suburban WWTP	4.3	3.5	3.1	2.8
Lost River Diversion Channel[1]	319	93	95	46
Columbia Forest Products	0.0005	0.0002	0.0001	0.0001
Klamath Straits Drain	155	82	94	86
Outflows (ft^3/s)				
Lost River Diversion Channel[1]	29	74	75	84
North Canal	47	55	48	52
Ady Canal	111	122	129	106
Keno Dam	2,246	1,071	1,095	907

[1]Lost River Diversion Channel is either an inflow or an outflow depending on the time of year.

comparing measured data at multiple locations throughout the reach to model predictions at the same date, time, and location. The model was initially calibrated for April through November 2007 and 2008, as those periods had the most extensive water-quality datasets available. Subsequently, the calibration period was extended to include calendar years 2006–09 to expand the range of hydrologic, water quality, and meteorological conditions that were modeled and to investigate the feasibility of modeling the reach with sparse datasets (2006 and 2009) as opposed to rich datasets (2007 and 2008). The CE-QUAL-W2 model uses a variable time step; for these models, the time step generally was between 50 and 800 seconds.

The model was set up to run in Pacific Standard Time (PST). Field data in Pacific Daylight Time (PDT) were converted to PST for model input files; model output files in PST were converted to PDT when necessary for comparison with field calibration data.

Model Grid

A CE-QUAL-W2 model grid is formed from model segments that connect together in the direction of flow. Each segment has layers of defined height that increase in width from the channel bottom to the top of the grid, thus resembling a cross-sectional shape with stacked rectangles. Segments are grouped together into "branches" that define specific river reaches, and branches can be grouped together as "waterbodies" that have similar meteorological conditions.

Cross-sectional shapes were extracted from a recent bathymetric survey of this reach (Eilers and Gubala, 2003) to produce the model grid using geographic information system (GIS) software. Segment boundaries were specified to form 102 active segments in the main river reach from Link River to Keno Dam (branch 1, fig. 2) and three segments describing a channel around an island in the Klamath River just upstream

of the inflow of the Lost River Diversion Channel (branch 2). Using GIS techniques, 10 equally spaced cross sections were subsampled within the designated segment boundary lines. The extracted cross-sectional shapes were averaged to determine a representative cross section for each model segment, recomputed to determine equivalent cell widths for a model grid with layer boundaries at specified elevations, and finally formatted as a CE-QUAL-W2 bathymetry input file. The model grid then was checked to ensure that layers at the river bottom were not advectively isolated (a minimum of two neighboring active cells are required in each layer to ensure that advective flow can occur) and that layer widths were greater than 5 m as recommended by the CE-QUAL-W2 development team (Cole and Wells, 2008). Adherence to these requirements is necessary to ensure that run times are reasonable and model instabilities are minimized. Segment lengths ranged between 477 and 1,170 ft (145.5 and 356.9 m) and averaged 1,009 ft (307.6 m); layer heights were all set to 2 ft (0.6096 m). A grid with a layer height of 1 ft (0.3048 m) also was tested, but results were not improved and run time was faster with 2-ft layers. The length of branch 1, from the end of Link River to Keno Dam, totaled 19.6 mi (31.5 km).

Model Data

Meteorology

The CE-QUAL-W2 model requires several meteorological data types as model input, including air temperature, dewpoint temperature, wind speed, wind direction, cloud cover, solar radiation, precipitation, and the temperature of precipitation. In a preliminary analysis, meteorological datasets were obtained and compared from several local stations: KLMT (Klamath Falls Airport), KFLO (Klamath Falls Agrimet), KENO (Reclamation), and SSH

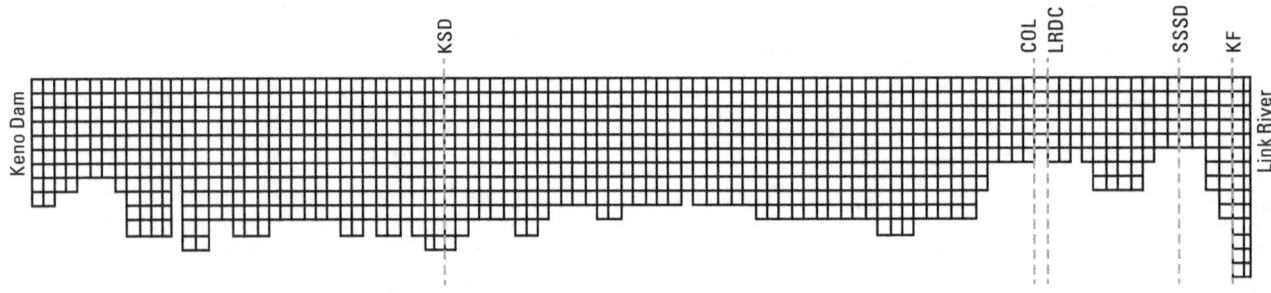

EXPLANATION

Vertical grid lines represent model segment boundaries
Horizontal grid lines represent boundaries between model layers
Each cell in the grid has a specific width from bank-to-bank and
    widths increase from the bottom of the grid to the top
    (bank-to-bank dimension not shown)
Dashed lines show the location of inflows

COL   Columbia Forest Products
KF   Klamath Falls wastewater treatment plant
KSD   Klamath Straits Drain
LRDC   Lost River Diversion Channel
SSSD   South Suburban wastewater treatment plant

Figure 2.   A longitudinal transect through the main branch (branch 1) of the model grid from the mouth of Link River (right) to Keno Dam (left), Oregon, representing 19.6 miles of the Klamath River.

(USGS South Shore of Upper Klamath Lake). Hourly data from KLMT were determined to be representative and used for most required meteorological inputs, except for solar radiation, which was obtained from KFLO. Cloud-cover data were obtained from the KLMT Automated Surface Observing System (ASOS) station and converted to model units. The ASOS laser beam ceilometer measured cloud cover up to 12,000 ft (National Oceanic and Atmospheric Administration, 1998). When the ASOS data reported clouds at multiple elevations, the elevation with the most cloud coverage was used for model input, similar to methods described by the U.S. Environmental Protection Agency (1997). None of the stations measured the temperature of precipitation directly; therefore, the precipitation temperature was assumed to be equal to air temperature when air temperature was above 0 °C. Precipitation temperature was set to 0 °C when air temperature was below 0 °C.

Riparian vegetation along this reach was dominated by low growing herbaceous annuals (for example, *Scirpus* sp.), and woody riparian vegetation was largely absent. Given the width of the river throughout this reach, vegetative riparian shading was deemed negligible and set to zero in the model. Similarly, the Link–Keno reach traverses a nearly flat plain with few nearby hills or mountains to provide any substantive topographic shading; shading from topographic features also was set to zero in the model.

Hydrology

The Klamath River begins at the downstream end of Link River, a short distance downstream of Upper Klamath Lake (fig. 1). The inflow from Link River forms the upstream boundary of the model. Streamflow in Link River upstream of the inflow from the Westside Power Canal was measured every 30 minutes by USGS gage 11507500 (fig. 1, table 2). When part of the flow was routed through the Westside Power Canal, those daily flow values, obtained from PacifiCorp, were added to the USGS gaged flows to provide the total Link River inflow to the Klamath River.

Several important tributary inflows and withdrawals were included in the model. Tributaries with measured daily flow data included the Klamath Falls WWTP, South Suburban WWTP, and Columbia Forest Products (all three report discharge information to the Oregon Department of Environmental Quality [ODEQ], and the Klamath Straits Drain and Lost River Diversion Channel (measured by Reclamation). Gaged withdrawals from this reach of the Klamath River include the Ady and North Canals, both with daily flow data from Reclamation.

Release rates through Keno Dam were derived from 30-minute flow data from USGS gage 11509500, about 1.5 mi downstream of the dam (fig. 1, table 2). Releases occur through three structures: spill gates, sluice conduit, and fish ladder. Accordingly, the model outflow was apportioned into these three outflows at Keno Dam. The sluice conduit and fish ladder flows were about 130 and 70 ft³/s, respectively,

whereas the spill gates released additional water when flow was greater than 200 ft³/s (PacifiCorp, 2002). The highest mean flows were from 2006 and the lowest mean flows were from 2009 for the 2006–09 modeling period (table 1). Daily average water-surface elevations measured at Keno Dam were provided by PacifiCorp and were used in the water balance.

Water Temperature and Water Quality

Water-quality measurements within the reach were made using three methods: (1) continuous monitors that were deployed at a fixed depth (every 30–60 minutes), (2) vertical profile measurements at multiple depths through the water column (weekly), and (3) grab samples at specific locations and depths at regular intervals (weekly, monthly).

Continuous Measurements and Vertical Profiles

Reclamation staff deployed and maintained 11 water-quality monitors in the Link–Keno reach during 2006–09 (table 2). The monitors provided hourly measurements throughout the year of water temperature, pH, dissolved oxygen, specific conductance, and sensor depth. Monitors were exchanged weekly from April through October with freshly cleaned and calibrated instruments; from November through March, the monitors were exchanged every 2 weeks. Vertical profiles of these parameters and turbidity as well as Secchi depth measurements were collected during site visits. An overlap period was implemented in the spring of 2007, such that when the clean and freshly calibrated instrument was deployed, the original instrument was left in the river for an additional day. This overlap period was useful during processing of the data to assure data continuity during monitor exchange. Monitor data for calendar years 2006–09 were loaded into the USGS National Water Information System (NWIS) database and processed to correct for fouling and instrument drift using USGS protocols and methods (methods modified from Wagner and others, 2006). Water temperature also was measured every 30 minutes at two USGS sites, one in Link River and the other below Keno Dam (table 2).

The continuous monitor dataset was extensive, although some minor gaps were present in the dataset as a result of site visits, probe malfunction, or biological interference (such as snails getting into the conductance probes). For the purpose of creating model boundary inputs, short gaps in the data on the order of a few hours were filled by linear interpolation using data before and after the gap. (The original corrected datasets in NWIS were not modified in this manner.) Longer data gaps, on the order of several days to a week, were filled by comparing (through plots and regressions) daily and seasonal patterns from existing datasets, then choosing an appropriate gap-filling method. For example, linear regression between measured values on each end of a gap could be used when little daily variation was noted for a constituent at a site or time period. However, summer water temperature and dissolved oxygen had strong daily cycles that required other gap-filling methods. Those gaps were filled either by using data directly from a nearby site when justified by close proximity (for example, water temperature gaps at site 11507501 were filled with hourly measurements from site 11507500, or by using modified data from a nearby site with similar cycles, shifted up or down if necessary to smoothly fill the gap.

The continuous monitor at the Lost River Diversion Channel began operation in early April 2007; therefore, estimates for water temperature, dissolved oxygen, and specific conductance for this site are more uncertain for January 2006 through March 2007. For the model boundary condition inputs prior to April 2007, water temperature at Lost River Diversion Channel was estimated by using measured temperature from the Klamath Straits Drain for periods when water temperature was less than 10 °C, and from Link River for periods when water temperature was greater than 10 °C. Lost River Diversion Channel specific conductance was dependent on whether flow was to or from the Klamath River, so longer data gaps were filled by assigning a specific conductance of 123 µS/cm for periods when flow was from the Klamath River and a value of 362 µS/cm for periods when flow was to the Klamath River. Dissolved-oxygen concentrations in Lost River Diversion Channel were estimated using several point measurements from upstream in the Lost River at Wilson Reservoir, and then incorporating and shifting dissolved-oxygen data from other sites with a similar magnitude of daily concentration cycles.

Table 2. Location of continuous-flow and water-quality monitors used for the development of model input or calibration of water temperature, dissolved oxygen, and specific conductance in the upper Klamath River, Oregon.

[Data collected during site visits to maintain the monitors, including turbidity, algal density observations, and vertical profiles, also were used for model calibration. Most monitors are owned and maintained by Bureau of Reclamation except for sites 11507500 and 11509500, which are USGS sites. **Site name:** RR, railroad. **Constituents:** DO, dissolved oxygen, Q, flow; SC, specific conductance; T, water temperature. **Use:** C, calibration; I, input. **Abbreviation:** m, meter]

Site name	Site identification No.	Latitude and longitude	Sensor depth (m)	Constituents	Use
Link River at Klamath Falls	11507500	42° 13' 10" -121° 47' 25"		Q, T	I
Link River below Keno Canal	11507501	42° 13' 10" -121° 47' 25"	1.0	DO, pH, SC, T	I
Klamath River at RR Bridge, at Lake Ewauna [top]	421209121463000	42° 12' 09" -121° 46' 30"	1.0	DO, pH, SC, T	C
Klamath River at RR Bridge, at Lake Ewauna [bottom]	421209121463001	42° 12' 09" -121° 46' 30"	2.4	DO, pH, SC, T	C
Lost River Diversion Channel near Klamath River	421015121471800	42° 10' 15" -121° 47' 18"	1.0	DO, pH, SC, T	I
Klamath River at Miller Island Boat Ramp [top]	420853121505500	42° 08' 53" -121° 50' 55"	1.0	DO, pH, SC, T	C
Klamath River at Miller Island Boat Ramp [bottom]	420853121505501	42° 08' 53" -121° 50' 55"	3.8	DO, pH, SC, T	C
Klamath Straits Drain near Highway 97	420451121510000	42° 04' 51" -121° 51' 00"	1.0	DO, pH, SC, T	I
Klamath River at Site KRS12a, near Rock Quarry [top]	420615121533600	42° 06' 15" -121° 53' 36"	1.0	DO, pH, SC, T	C
Klamath River at Site KRS12a, near Rock Quarry [bottom]	420615121533601	42° 06' 15" -121° 53' 36"	3.9	DO, pH, SC, T	C
Klamath River above Keno Dam, near Keno[top]	11509370	42° 07' 41" -121° 55' 44"	0.9	DO, pH, SC, T	C
Klamath River above Keno Dam, near Keno [bottom]	420741121554001	42° 07' 41" -121° 55' 44"	3.8	DO, pH, SC, T	C
Klamath River below Keno Dam, at Keno	11509500	42° 08' 00" -121° 57' 40"		Q, T	I (Q) C (T)

Grab Samples

Extensive monitoring of water quality with grab samples was conducted as part of the study from April through November of 2007 and 2008. Samples were collected weekly at five mainstem sites and from major tributaries. These intensive sites are designated as both USGS and Reclamation in table 3. Samples were analyzed for total nitrogen and phosphorus; particulate carbon and nitrogen; filtered orthophosphate, nitrite, nitrite plus nitrate, ammonia, and organic carbon; and the enumeration and species identification of phytoplankton and zooplankton. Methods and results for this data-collection effort were documented by Sullivan and others (2008, 2009). More limited pilot sampling also was conducted in late August to early September 2006.

Water-quality data for the releases from the Klamath Falls and South Suburban WWTPs and from Columbia Forest Products were obtained from discharge monitoring reports (DMR) provided to ODEQ. Alkalinity was not measured at any of the point sources; it was estimated to be equal to the alkalinity in Link River.

Other datasets were required to construct model boundary conditions and calibration datasets for time periods outside of the intensive sampling efforts in 2007 and 2008 (January 2006–March 2007, December 2007–March 2008, and December 2008–December 2009). Water-quality data from Reclamation, ODEQ, and PacifiCorp were obtained and compiled to fill these gaps (table 3). If multiple sources of data were available for a site, the results were plotted together and examined. If the data from one source had notably different concentrations or temporal patterns than data from other sources for the same time and location, this difference was investigated further. If a sampling problem was documented or if no quality-assurance or quality-control data were available, the data were not used for this study.

Some of the grab sample data had been censored (presented as less than a reporting level), and special methods were used to address the issues associated with these data. Laboratories usually do not report raw results if the values are below a certain reporting level; instead, that result is simply reported to be less than the reporting level. Although such procedures are useful for avoiding certain types of errors, an uncensored value is required for model input. The raw (uncensored) laboratory data were obtained and used to create the model input file for periods with data from the USGS National Water Quality Laboratory (April–November 2007 and 2008), when data were censored at the reporting level. Values from other agencies reported as less than the reporting level were set to one-half of the reporting level.

Acoustic Doppler Current Profiler Data

Measurements of water velocity (speed and direction) were made with acoustic Doppler current profilers (ADCPs) at multiple sites in the Link–Keno reach and as part of an intensive examination of circulation patterns in the Lake Ewauna reach of the Klamath River. An ADCP transmits a fixed frequency sound pulse and measures the frequency shift of acoustic echoes of suspended particles. A series of cross-sectional measurements at selected locations between Link River and Keno Dam were completed using an ADCP from a moving boat on May 30 and September 29, 2007. These measurements included data from most of the cross-sectional area, although a portion near the riverbanks was not measurable because of minimum depth requirements. Portions near the surface and bottom of the water column were not measurable due to the ADCP depth into the water column, blanking distance (a blind zone immediately beneath the ADCP), and interference near the river bottom. Methods and complete results for the cross-sectional work were documented by Sullivan and others (2008). ADCPs were fixed on the river bottom in 2008 and oriented vertically at three sites to measure water velocity profiles continuously from summer through fall. Similar to the cross-sectional work, a portion near the surface and bottom of the water column could not be measured due to signal interference near the water surface, blanking distance, and the height of the instrument mount above the river bottom.

One ADCP was deployed on the river bottom at Keno from June 5 through December 7, 2008. That instrument (Argonaut-XR) used three acoustic beams equally spaced at 120°, with each beam 25° from the vertical instrument axis. Water velocity (speed and direction) through the water column was measured every 30 minutes from June 5 to September 23 and every 10 minutes from September 23 to December 7. Water speed and direction data were processed and lumped into four layers of the water column, each 1.4 ft (0.43 m) high. Two other ADCPs (EWA1, EWA2) were deployed in Lake Ewauna from June 4 through October 10, 2008. Those instruments (1,200 kHz Teledyne RD Workhorse) used four acoustic beams equally spaced at 90°, with each beam 20° from the vertical instrument axis. Velocity through the water column was measured every 30 minutes at both Lake Ewauna sites. The morphology of Lake Ewauna is variable, with a relatively deep channel to the west and a wide, shallow area to the east. Thus, the ADCP on the west (EWA1) measured velocity in a deeper water column, usually with 11 data "bins" with a height of 0.7 ft (0.2 m). The ADCP deployed to the east (EWA2) was in the shallow area, resulting typically in three data bins with a height of 0.7 ft. Measured directions for all velocity data were corrected for magnetic declination and are reported relative to true north.

Table 3.   Location of water quality grab sample data used to support model development or calibration for the upper Klamath River, Oregon.

[**Site name:** RR, railroad. **Source:** ODEQ, Oregon Department of Environmental Quality; PC, PacifiCorp; Reclamation, Bureau of Reclamation; USGS, U.S. Geological Survey. **Use:** C, calibration; I, input]

Site name	Source	Site identification No.	Latitude and longitude or Klamath river mile	Use
Link River below Keno Canal	USGS Reclamation	11507501	42° 13' 10" -121° 47' 25"	I
Link River at mouth	ODEQ	10768	42° 13' 08" -121° 47' 18"	I
Link River	PC	KR25312	253.12	I
Klamath Falls wastewater treatment plant outfall	ODEQ	13174	42° 12'' 57" -121° 46' 36"	I
South Suburban Sanitation District outfall	ODEQ	13316	42° 11' 51" 121° 46' 13"	I
Klamath River at RR Bridge, at Lake Ewauna [top]	USGS Reclamation	421209121463000	42° 12' 09" -121° 46' 30"	C
Klamath River at RR Bridge, at Lake Ewauna [bottom]	USGS Reclamation	421209121463001	42° 12' 09" -121° 46' 30"	C
Lost River Diversion Channel near Klamath River	USGS Reclamation	421015121471800	42° 10' 15" -121° 47' 18"	I
Lost River Diversion Dam	USBR	K-5	42° 09' 18" -121° 39' 46"	I
Columbia Forest Products outfall	ODEQ	789	42° 10' 40" -121° 47' 57"	I
Klamath River at Miller Island Boat Ramp [top]	USGS Reclamation	420853121505500	42° 08' 53" -121° 50' 55"	C
Klamath River at Miller Island Boat Ramp [bottom]	USGS Reclamation	420853121505501	42° 08' 53" -121° 50' 55"	C
Klamath Straits Drain near Highway 97	USGS Reclamation	420451121510000	42° 04' 51" -121° 51' 00"	I
Klamath Straits Drain at Highway 97	USBR	K-1	42° 04' 51" -121° 50' 44"	I
Klamath Strait at Reclamation Pump Station F	ODEQ	10763	42° 04' 48" -121 50' 27"	I
Klamath River at Site KRS12a, near Rock Quarry [top]	USGS Reclamation	420615121533600	42° 06' 15" -121° 53' 36"	C
Klamath River at Site KRS12a, near Rock Quarry [bottom]	USGS Reclamation	420615121533601	42° 06' 15" -121° 53' 36"	C
Klamath River above Keno Dam, near Keno [top]	USGS Reclamation	11509370	42° 07' 41" -121° 55' 44"	C
Klamath River above Keno Dam	ODEQ	10765	42° 07' 41" -121° 55' 44"	C
Klamath River above Keno Dam, near Keno [bottom]	USGS Reclamation	420741121554001	42° 07' 41" -121° 55' 44"	C
Klamath River below Keno Dam, at Keno	USGS	11509500	42° 08' 00" -121° 57' 40"	C
Klamath River below Keno Dam	PC	KR23334	233.34	C

Model Development

Water Balance

Results from initial model runs that included precipitation, evaporation, and measured inflows and outflows identified differences between modeled and measured water-surface elevations. This typically occurs due to one or more of several reasons, including the presence of ungaged surface-water inflows or outflows, groundwater sources or sinks, or error in the measurement of gaged inflows and outflows. The sum of these ungaged flows and gage errors must be quantified and accounted for to ensure that the model accurately simulates the stage, velocity, overall flow, and travel time through the model domain.

The CE-QUAL-W2 model includes an option to apply a "distributed tributary" input that accounts for ungaged water inputs or withdrawals. The addition of a distributed tributary is a common means of completing the water balance (Cole and Wells, 2008). A distributed tributary specifies the total amount of water gained or lost along an entire model branch, instead of only at one location (as is the case with a normal tributary). Distributed tributary flows for these models were computed and applied at daily intervals and could be either positive (representing the need for additional inflows) or negative (representing the need for additional outflows) at different times of the year. The total flow associated with the distributed tributary was small relative to total inflows and outflows; it made up 3, 1, 5, and 2 percent of total inflows and 4, 4, 2, and 3 percent of total outflows in 2006, 2007, 2008, and 2009, respectively. The final modeled water-surface elevations, including the distributed tributary input, are shown with measured values in figure 3.

Water Temperature Boundary Conditions

Water temperature was measured at most boundary and tributary locations on an hourly basis, and those data were used directly as model input. Water temperature showed a strong seasonal pattern for most inflows, with winter temperatures approaching 0 °C and maximum summer temperatures in July and August between 20 and 28 °C. The Klamath Falls WWTP effluent was warmer than the Klamath River in winter, with temperatures rarely dipping below 10 °C. Water temperature was not reported for the Columbia Forest Products outfall; its temperature was assumed to be similar to that of the South Suburban WWTP because both sources discharge from holding ponds.

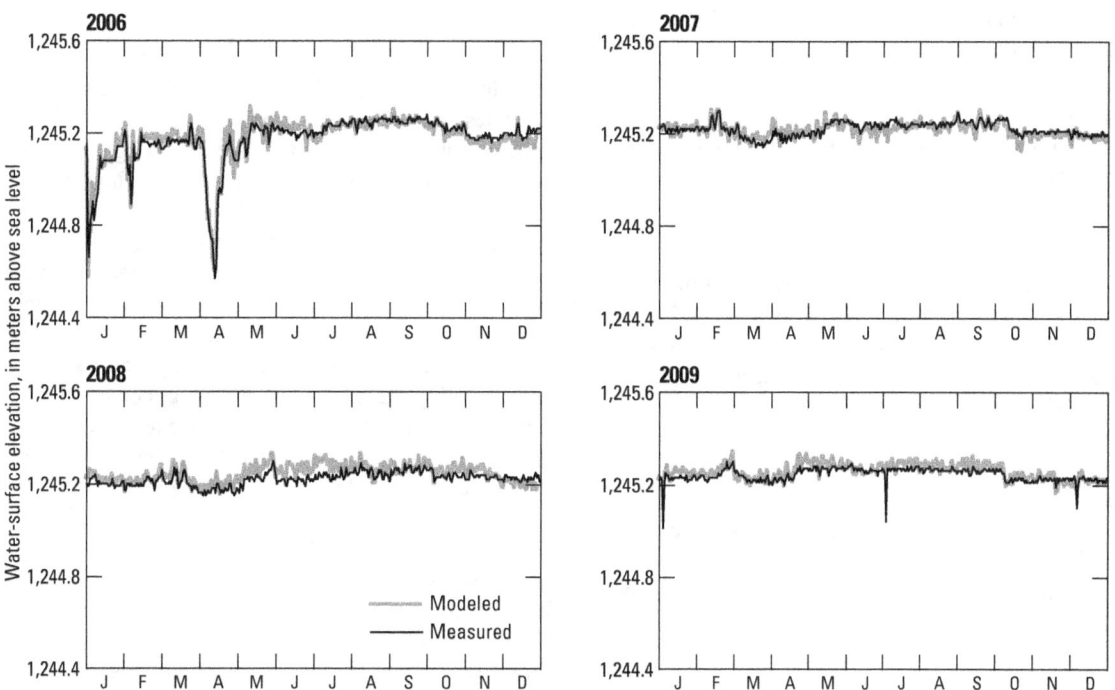

Figure 3. Daily average measured and modeled water-surface elevations at Keno Dam forebay, Oregon.

The distributed tributary probably represents surface water more than groundwater. Groundwater discharge to the Klamath River between Link River and Keno Dam has been estimated to be minor (Gannett and others, 2007). Furthermore, the measurement errors associated with flow data at Reclamation canal sites may be substantial, perhaps greater than 15 percent (Risley and others, 2006). Therefore, it is likely that the water accounted for by the model's distributed tributary is mostly surface water; accordingly, the distributed tributary flows were estimated to be 90 percent surface water and 10 percent groundwater. Surface-water temperatures were estimated by using those measured in the Klamath Straits Drain. Groundwater temperature was estimated to be 14 °C, from a groundwater temperature contour map of the area (Sammel, 1980) as well as from nearby well water temperatures measured during 1995–2008 and stored in the ODEQ LASAR database (Oregon Department of Environmental Quality, 2009).

Water Quality Boundary Conditions

For many water-quality constituents, measured data were used directly as model input, including data for orthophosphate, ammonia, nitrate plus nitrite, and dissolved-oxygen concentrations. Because the various constituents were collected at different intervals (weekly, hourly), and needed to be at the same interval in the water quality input files, linear interpolation between some data points was used to produce model input files with a consistent time interval. Some data processing was required for other modeled constituents. Further details for these constituents are provided in the next section of this report.

Total Dissolved Solids

Total dissolved solids (TDS) are a measure of the sum of dissolved constituents in water. The CE-QUAL-W2 model uses TDS to calculate water density and ionic strength, so it is important in determining the vertical placement and subsequent mixing of tributary inflows. TDS is not routinely measured in the Klamath River, but it is closely related to specific conductance, which is a measure of the electrical conductance of water and a gross estimate of the amount of dissolved ions in solution. Hourly or daily measurements of specific conductance were available for most sites. Specific conductance and TDS at many sites can be described by the following equation from Hem (1985):

$$TDS = A \times specific\ conductance,\qquad(1)$$

where

A is a factor between 0.54 and 0.96, but has a range of 0.55 to 0.75 for most natural waters.

To derive an estimate of A, data from Klamath River samples collected in this reach and stored in the USGS NWIS database were examined for samples in which both TDS, from residue on evaporation at 180 °C, and specific conductance were measured. Most of the 18 samples collected in 1961–62 had calculated values of A from 0.61 to 1.11. No regular spatial or temporal patterns were evident in those data; therefore, the median value 0.69 was chosen as the value for A in equation 1 to derive TDS inputs from hourly or daily specific conductance. Data for 38 groundwater samples collected between 1995 and 2008 near the river (Oregon Department of Environmental Quality, 2009) for which both TDS and specific conductance were measured also resulted in a median value of 0.69 for A, with a range between 0.57 and 0.83. Equation 1 with $A = 0.69$ was used to generate TDS data from specific conductance measurements at Link River, the Lost River Diversion Channel, the Klamath Straits Drain, and the Klamath Falls WWTP. Only two measurements of specific conductance were available for the effluent of the South Suburban WWTP. Therefore, those measurements were averaged and converted to a constant TDS for that point source. Columbia Forest Products had no existing specific conductance or TDS data; an estimate of 500 mg/L was used in the model. Equation 1 was used in reverse to convert modeled TDS to specific conductance for comparison to measured data during the calibration process.

Inorganic Suspended Sediment

Inorganic suspended sediment (ISS) consists of mineral particles, such as clays, suspended in the water column. This constituent influences water density and light penetration in the model. Direct analyses of ISS were infrequent, but could be correlated to more frequently measured turbidity measurements to estimate ISS concentrations for the model inputs (fig. 4). Turbidity data were collected in triplicate by the Reclamation field crew during trips to maintain the continuous water-quality monitors, resulting in approximately weekly measurements throughout the year. ISS is determined by measuring the difference in the mass on a filter before and after baking it in a muffle furnace to drive off organic matter. This is similar to subtracting measurements of volatile suspended solids (VSS) from measurements of total suspended solids (TSS). Turbidity is an optical property of water that measures scattering and absorption of light due to suspended and dissolved material, including ISS, algae, and particulate and dissolved organic matter (POM and DOM); but, use of the near-infrared wavelength by turbidity instruments, such as those used to collect this dataset, can help exclude the effect of DOM. Summer samples were excluded from the correlation because algae and POM concentrations are highest during this season; samples were included in the regression only when the qualitative visual algal observation was 0. The relation used to develop ISS inputs was:

$$ISS = 0.1776 \times Turbidity^{1.5505}.\qquad(2)$$

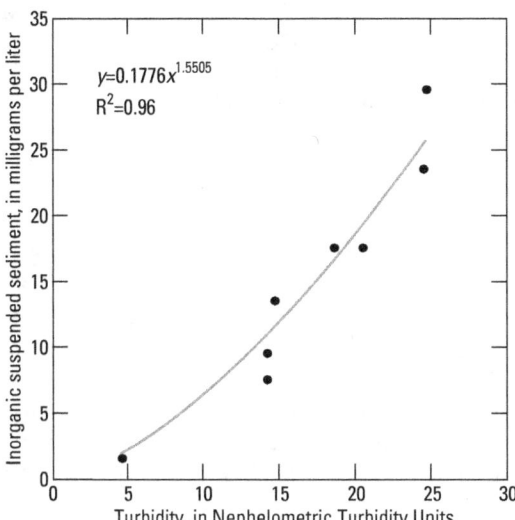

Figure 4. Regression between measured turbidity and inorganic suspended sediment concentration when algae were not observed to be present (field ALGAE observations = 0) in the upper Klamath River, Oregon. Inorganic suspended sediment is calculated by subtracting volatile suspended sediments (the organic fraction) from total suspended sediment. Samples collected by the Bureau of Reclamation, 2009–10.

The highest measured turbidity used to develop this correlation was <30 nephelometric turbidity units (NTU), as VSS and TSS data were not available for higher turbidities. Greater uncertainties are present when using this equation with turbidities higher than 30 NTU; high turbidities were observed in winter in some tributaries (up to several hundred NTU) but these occurrences were infrequent. Summer ISS concentrations for the Link River, Lost River Diversion Channel, and Klamath Straits Drain inflows were set at approximately 5.0 mg/L. Only TSS (not ISS) data existed for the NPDES-regulated point sources. With the assumption that most of those suspended materials would be of organic origin, the ISS for the NPDES point sources was estimated to be 0 mg/L.

Algae

The CE-QUAL-W2 model allows the simulation of multiple algal groups, limited principally by the availability of supporting data. Data on the amount and species composition of algae in the Klamath River were available for April–November 2007 and 2008 (Sullivan and others, 2008, 2009). Examination of the spatial and temporal patterns in the algal species and biovolume data led to a logical and data-supported separation of the Klamath River algae community into three groups for this model:

1. Blue-green algae,

2. Diatoms, and

3. Other algae.

The blue-green algae, also known as cyanobacteria, dominate the assemblage in summer and consist largely of AFA with less frequent occurrence of *Anabaena flos aquae*. Diatoms are common in the Klamath River upstream of Keno Dam during spring, with species such as *Fragilaria capucina* var. *mesolepta*, *Stephanodiscus astraea* var. *minutula*, *Asterionella formosa*, and *Melosira granulata*. The "other" algae group from the Klamath River data includes species such as *Chlamydomonas* sp. and *Cryptomonas erosa*.

The algae species data for the Klamath River are reported as cell density (number per milliliter of sample) or biovolume (cubic microns per milliliter) estimates, but the model requires input in units of dry weight concentration (gram per cubic meter or milligrams per liter). Algal carbon was calculated from biovolume by using an equation from Rocha and Duncan (1985). This equation was developed by compiling results from 47 studies of various freshwater algal species that reported both cell volumes and cell carbon concentrations. Once the algal carbon concentration was obtained, the algal carbon to organic matter ratio was applied, of which the derivation is described below.

Some required model inputs, including algal and particulate organic matter concentrations, were not directly quantified at inflow sites in winter and during 2006 and 2009. These input concentrations were estimated by using qualitative observations, seasonal concentration patterns from other years, or analysis of trends in data at nearby sites. For example, blue-green algae concentrations were important to include in the models because they are critical to the water quality in the Link–Keno reach. Qualitative algal density observations from Reclamation were used to develop quantitative estimates to fill data gaps. The Reclamation field crew visually inspected the water for the presence of algae during each field site visit, and recorded the visual observation as a number from 0 to 5, ranging from "not visible" (0) to "dense" (5). These observations were made approximately weekly during site visits for grab sampling, sonde deployment, and sonde retrieval. These visual observations would note the presence of large algae such as AFA, but are not generally indicative of less visible algae such as diatoms. Comparisons to measured algal biovolume and species data were made for dates and sites where data were available (fig. 5). Directly measured blue-green algal biovolume and species data were used first for model input; estimates from field observed algal density were then used to fill out input datasets for periods with no other algal data. These estimates were used for 2006 for Link River and Klamath Straits Drain; 2009 for Link River, Lost River Diversion Channel, and Klamath Straits Drain; 2007 for Klamath Straits Drain and Lost River Diversion Channel, and part of 2008 for Lost River Diversion Channel. As a secondary source of algal information when direct sample

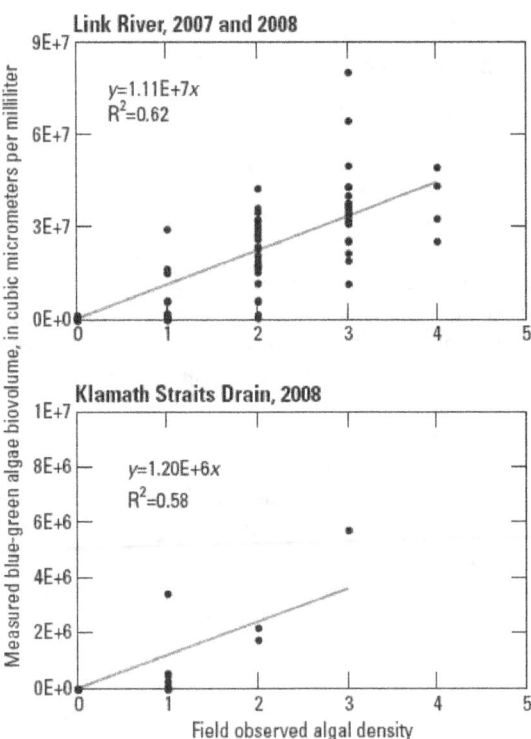

Figure 5. Regression between measured blue-green algae biovolume and field observed algal density at the Link River (2007 and 2008 data) and Klamath Straits Drain (2008 data) sites, Oregon.

data did not exist, the seasonal patterns in the algae input datasets were compared to chlorophyll *a* data measured by the Klamath Tribes at the Pelican Marina site in Upper Klamath Lake in the vicinity of Link Dam (fig. 1). Minor modifications to the seasonal patterns in the estimated Link River datasets for 2006 and 2009 were made after this comparison from the added knowledge about temporal patterns in algae population dynamics for short periods without direct algal data.

Organic Matter

Organic matter in CE-QUAL-W2 is separated into four compartments based on relative rates of decay (fast [labile] or slow [refractory]) and its physical status (particulate or dissolved). Dissolved and particulate organic matter (DOM and POM, respectively) are designated as labile (L prefix) or refractory (R prefix). Labile dissolved organic matter, for example, is designated as LDOM. The process of converting laboratory measurements into concentrations of LDOM, RDOM, LPOM, and RPOM for model input required several assumptions and insights based on data and experiments.

To determine the concentration of DOM, measured dissolved organic carbon concentrations were converted using the organic carbon to organic matter ratio (derivation described below). In summer, Link River and Lost River Diversion Channel DOM were partitioned into 8 percent

LDOM and 92 percent RDOM; Klamath Straits Drain was partitioned into 5 percent LDOM, and 95 percent RDOM. The LDOM for these three inputs was lowered to 1 percent in winter. These partitionings were based on general knowledge that DOM tends to be refractory and specific knowledge from biochemical oxygen demand (BOD) experiments showing DOM to be refractory in the Upper Klamath River (Sullivan and others, 2010). The data also showed a seasonal component for decay rates for this mostly refractory material, with higher lability in summer. Klamath Falls and South Suburban WWTP inflows to the Klamath River were set at 10 and 15 percent LDOM, respectively, throughout the year. The Columbia outfall DOM was partitioned to 95 percent LDOM to approximate the measured BOD values. These point sources were not assigned a seasonal component because the source material was assumed to stay fairly constant through the year.

To determine the concentration of POM for Link River, Lost River Diversion Channel, and Klamath Straits Drain, the concentration of live algae was subtracted from the total particulate organic matter concentration, which was derived from measured particulate carbon and the organic carbon to organic matter ratio. In summer, Link River and Lost River Diversion Channel POM was partitioned into 80 percent LPOM and 20 percent RPOM, and Klamath Straits Drain POM was partitioned into 50 percent LPOM and 50 percent RPOM. In winter, the fraction of LPOM for all three of these inputs was lowered to 1 percent. These partitionings are based on BOD experiments that showed summer algal-derived particulate matter in the Link–Keno reach decomposed quickly (Sullivan and others, 2010). Whereas algae, in general, are a source of labile organic matter, blue-green algae are an especially labile form of algae, perhaps because they lack a cellulose wall (Boers and others, 1991). The BOD experiments showed less lability in Klamath Straits Drain, probably due to a larger variety of algal species and organic matter from a different source than Upper Klamath Lake. Fewer data were available to characterize the concentration and lability of organic matter in winter. Because the large algal blooms do not occur in winter, it was assumed that winter particulate organic matter was sourced more from terrestrial plant detritus or decaying macrophytes, older material which is rich in cellulose and lignin and much less labile than blue-green algae. Klamath Falls WWTP and South Suburban WWTP inflows were set at 65 percent LPOM throughout the year; measured BOD was used to estimate the labile-refractory organic matter split for those inflows. The Columbia outfall POM was partitioned to 95 percent LPOM to approximate the measured BOD values.

Abundance data for bacteria and zooplankton in this reach of the Klamath River provide useful insights and checks on the ranges of the estimated algal and POM data. Although bacteria and zooplankton are not explicitly represented in this model (they are part of the particulate organic matter fraction), the bacteria and zooplankton measurements were converted to carbon concentration to compare to particulate organic carbon and algal carbon. Bacteria biovolumes were converted to bacterial carbon using a conversion factor from Nagata (1986). Zooplankton densities were converted to zooplankton carbon using equations from the U.S. Environmental Protection Agency (2003) and ratios of body length to dry body weight (Allan Vogel, ZP's Taxonomic Services, written commun., November 2009); the dry body weights were converted to carbon using a dry weight:wet weight conversion factor and an estimate of zooplankton carbon per unit biovolume (Latja and Salonen, 1978). These calculations show that in summer, depending on the date, either algal biovolume or nonliving POM (dead algae, detritus) made up most of the carbon biomass in the water column at the Link River site, followed by zooplankton and bacteria. The predominance of algae as a proportion of total particulate organic matter biomass in this predominantly eutrophic system is similar to the biomass partitioning observed by Auer and others (2004) in a study of 55 shallow mesotrophic to eutrophic lakes; with increasing eutrophication, phytoplankton, especially blue-green algae, made up a greater proportion of water column particulate biomass compared to bacteria and zooplankton.

Distributed Tributary Water Quality

For water temperature, the distributed tributary was estimated to be a mixture of 90 percent surface water and 10 percent groundwater. Groundwater nitrogen, phosphorus, carbon, and oxygen were estimated by examining groundwater data collected by ODEQ from 1995 to 2008 at nonlandfill wells near the upper Klamath River (Oregon Department of Environmental Quality, 2009). Algae and ISS were set to 0 for the groundwater part of the distributed tributary inflow. The surface-water part of the distributed tributary was assumed to have a TDS equal to that from the Klamath Straits Drain. Groundwater TDS concentrations were estimated as a constant 300 mg/L by examining the groundwater specific conductance contour map of the area by Sammel (1980) and from measured TDS values from 1995 to 2008 (or specific conductance converted to TDS) in nearby wells (Oregon Department of Environmental Quality, 2009).

Water-Quality Parameters

CE-QUAL-W2 requires a large number of model parameters (tables 4 and 5) that are used in the numerical representation of water-quality processes. These include rate constants, element ratios (stoichiometry), and a wide range of coefficients, half-saturation constants, and other parameters that specify the implementation details of the model algorithms. To construct an accurate and predictive model, parameters should be based on measurements or literature values, if available. The derivation of certain model parameters from data or the literature for the Link–Keno reach of the Klamath River is described in the following sections. Parameters for which there were limited information, or which may vary over a defined range, were adjusted during model calibration. The response of the model to selected parameters, especially those not based on data or literature values, was examined in sensitivity tests to determine their relative importance.

Algal and Organic Matter Stoichiometry

The model relies on a set of stoichiometric parameters that define the dry mass ratios of carbon, nitrogen, and phosphorus to total organic matter in algae and organic matter. Default model values for AC, AN, AP (the algal ratios for carbon, nitrogen, and phosphorus) and ORGC, ORGN, and ORGP (the organic matter ratios for carbon, nitrogen, and phosphorus) are 0.45, 0.08, and 0.005, respectively, such that the algal ratios are identical to the organic matter ratios. The parameter values used in this model (tables 4 and 5) were based on values from data, which were then adjusted during model calibration. Two datasets were used: (1) particulate organic carbon, particulate organic nitrogen, and particulate organic phosphorus collected at six samplings between mid-July and late September 2008 at Link River and Railroad Bridge by Watercourse Engineering, Inc. (Deas and Vaughn, 2011); and (2) particulate carbon and nitrogen samples collected weekly at a number of sites between April and November of 2007 and 2008 by Reclamation and analyzed by the USGS (Sullivan and others, 2008, 2009). These particulate data were derived from measurements of the composition of material captured from a water sample on a glass fiber filter (0.7 µm pore size), which would include contributions from algae, zooplankton, and particulate organic matter. The latter dataset showed that there was seasonality to the particulate C:N ratio in the Klamath River, with lower values in summer. This seasonality could have resulted from the presence of nitrogen-fixing blue-green algae in summer, which allows atmospheric nitrogen to be an additional source of nitrogen to that algal group. However, the model does not allow seasonal variation for these stoichiometric ratios. These ratios were adjusted during calibration and final values of AC, AN, AP (and ORGC, ORGN, and ORGP) were 0.46, 0.059, and 0.004, respectively. On a molar basis, this provides a 33:1 N:P stoichiometry, which is within the range of values measured by Deas and Vaughn (2011).

In addition to the ratios of C, N, and P to organic matter, the CE-QUAL-W2 model requires a value for the mass ratio between algal biomass and chlorophyll *a*. This ratio is used to convert model output into chlorophyll *a* concentrations for comparison to measured data. Although constant in the model, this ratio has been reported to be extremely variable, tied to light conditions and nutrient supply (Reynolds, 2006). Ratios of algal biomass to chlorophyll *a* were calculated using measured chlorophyll *a* data and algal biomass estimates from the Klamath River at Railroad Bridge, Miller Island, KRS12a, and Keno. Bearing in mind the inherent variability, a median calculated value of 0.031 mg algae per µg chlorophyll *a* was used in the model.

Table 4.   Model parameters used in the model for the Klamath River upstream of Keno Dam, Oregon.

[**Abbreviations:** C, degrees Celsius; g, gram; g/m^3, gram per cubic meter; g/m^2, gram per square meter; m, meter; 1/d, 1 per day; 1/m, per meter; m^2/g, square meter per gram; SOD, sediment oxygen demand; W, Watts; (W/m^2)/ C, Watt per square meter per degree Celsius]

Parameter	Value	Description
WSC	1.0	Wind sheltering coefficient, dimensionless
AFW	9.5	Coefficient in wind speed formulation
BFW	0.46	Coefficient in wind speed formulation
CFW	2.00	Coefficient in wind speed formulation
EXH2O	1.217	Light extinction coefficient for water and dissolved constituents, 1/m
EXSS	0.167	Light extinction due to inorganic suspended solids, m^2/g
EXOM	0.147	Light extinction due to organic suspended solids, m^2/g
BETA	0.45	Fraction of solar radiation absorbed at water surface, dimensionless
TSED	14.0	Sediment temperature, C
CBHE	0.30	Coefficient of bottom heat exchange, (W/m^2)/ C
LDOMDK	0.121	Labile dissolved organic matter decay rate, 1/d
RDOMDK	0.0005	Refractory dissolved organic matter decay rate, 1/d
LRDDK	0.002	Labile to refractory dissolved organic matter conversion rate, 1/d
LPOMDK	0.101	Labile particulate organic matter decay rate, 1/d
RPOMDK	0.0005	Refractory particulate organic matter decay rate, 1/d
LRPDK	0.002	Labile to refractory particulate organic matter conversion rate, 1/d
POMS	0.25	Particulate organic matter settling rate, m/d
OMT1	2.0	Lower temperature parameter for organic matter decay, C
OMT2	25.0	Upper temperature parameter for organic matter decay, C
OMK1	0.15	Fraction of organic matter decay rate at OMT1
OMK2	0.90	Fraction of organic matter decay rate at OMT2
ORGP	0.004	Stoichiometric equivalent between organic matter and phosphorus, g P/g OM
ORGN	0.059	Stoichiometric equivalent between organic matter and nitrogen, g N/ g OM
ORGC	0.46	Stoichiometric equivalent between organic matter and carbon, g C/ g OM
PARTP	0.0	Phosphorus partitioning coefficient for suspended solids, dimensionless
PO4R	0.00208	Release rate of phosphorus from sediment, as a fraction of SOD
NH4R	0.002	Release rate of ammonium, as a fraction of SOD
NH4DK	0.0508	Ammonia nitrification rate, 1/d
NH4T1	4	Lower temperature parameter for ammonia nitrification, C
NH4T2	25	Upper temperature parameter for ammonia nitrification, C
NH4K1	0.1	Fraction of nitrification rate at NH4T1
NH4K2	0.99	Fraction of nitrification rate at NH4T2
NO3DK	2.60	Denitrification rate, 1/d
NO3S	0.01	Denitrification rate, loss to sediments, m/d
NO3T1	4.0	Lower temperature parameter for nitrate denitrification, C
NO3T2	20.0	Lower temperature parameter for nitrate denitrification, C
NO3K1	0.10	Fraction of denitrification rate at NO3T1
NO3K2	0.99	Fraction of denitrification rate at NO3T2
O2NH4	4.57	Oxygen stoichiometry for nitrification, g O$_2$/ g N
O2OM	1.4	Oxygen stoichiometry for organic matter decay, g O$_2$/ g OM
O2AR	1.1	Oxygen stoichiometry for algal respiration, g O$_2$/g algae
O2AG	1.4	Oxygen stoichiometry for algal primary production, g O$_2$/g algae
KDO	0.1	Dissolved oxygen concentration at which anaerobic processes are at 50 percent of maximum, g/m^3
SEDCI	8.0	Initial sediment concentration, g/m^2
SODT1	2.0	Lower temperature parameter for zero-order SOD or first-order sediment decay, C
SODT2	25.0	Upper temperature parameter for zero-order SOD or first-order sediment decay, C
SODK1	0.15	Fraction of SOD or sediment decay rate at SODT1
SODK2	0.90	Fraction of SOD or sediment decay rate at SODT2
SOD	2.3	Zero-order SOD for each segment, g O$_2$/ m^2/d
FSOD	0.101	Fraction of the zero-order SOD rate used
SSS	0.2	Inorganic suspended solids settling rate, m/d
C1	0.2	Reaeration equation coefficient
C2	0.05	Reaeration equation coefficient
C3	1.75	Reaeration equation coefficient

Table 5.   Algal parameters used in the model for the Klamath River upstream of Keno Dam, Oregon.

[**Abbreviations:** C, degrees Celsius; g, gram; g/g, gram per gram; g/m^3, gram per cubic meter; m, meter; m/d, meter per day; m^2/g, square meter per gram; W/m^2, Watt per square meter; 1/d, 1 per day]

Parameter	Blue-green algae	Diatoms	Other algae	Description
EXA	0.088	0.542	0.170	Light extinction due to algae, m^2/g
AG	3.09	0.80	1.4	Maximum algal growth rate, 1/d
AR	0.06	0.04	0.03	Maximum respiration rate, 1/d
AE	0.06	0.04	0.04	Maximum algal excretion rate, 1/d
AM	0.41	0.005	0.005	Maximum algal mortality rate, 1/ d
AS	0.80	0.01	0.005	Settling rate, m/d
AHSP	0.006	0.004	0.004	Algal half-saturation for phosphorus limited growth, g/m^3
AHSN	0.000	0.013	0.100	Algal half-saturation for nitrogen limited growth, g/m^3
ASAT	30	45	56	Light saturation intensity at maximum photosynthetic rate, W/m^2
AT1	12	4	12	Lower temperature parameter for rising rate function, °C
AT2	26	10	22	Upper temperature parameter for rising rate function, °C
AT3	35	16	25	Lower temperature parameter for falling rate function, °C
AT4	40	20	35	Upper temperature parameter for falling rate function, °C
AK1	0.1	0.1	0.1	Fraction of rate at AT1
AK2	0.99	0.99	0.99	Fraction of rate at AT2
AK3	0.99	0.99	0.99	Fraction of rate at AT3
AK4	0.1	0.1	0.1	Fraction of rate at AT4
AP	0.004	0.004	0.004	Stoichiometric equivalent between biomass and phosphorus, g/g
AN	0.059	0.059	0.059	Stoichiometric equivalent between biomass and nitrogen, g/g
AC	0.46	0.46	0.46	Stoichiometric equivalent between biomass and carbon, g/g
ACHLA	0.031	0.031	0.031	Ratio between algal biomass and chlorophyll a, milligram algae/µg chla
ALPOM	0.8	0.8	0.8	Fraction of algal biomass converted to particulate organic matter when algae die
ANPR	0.001	0.001	0.001	Algal half saturation preference constant for ammonium

Algal Rates and Coefficients

The CE-QUAL-W2 model simulates algal growth, respiration, excretion, mortality, and settling, and requires rates and coefficients, as well as half-saturation values for phosphorus- and nitrogen-limited growth and light saturation at the maximum photosynthetic rate. Each of the modeled algal groups requires values for these rates and coefficients as well as some parameters that define the temperature dependence of these processes (table 5). The rates used in the model are guided by values reported in the literature for similar types of algae, though rates can vary depending on a variety of factors; for example, growth rates are dependent on light, nutrient conditions, and water temperature at the time of measurement (Reynolds, 2006).

The rate of algal growth in the model is a maximum gross value not adjusted for losses such as respiration, mortality, excretion, or sinking. Algal growth rates in the model are temperature dependent. Temperature parameters were adjusted to define the blue-green and other algae as warm-water algae, and the diatoms as cool-water algae (see parameters AT1–AT4 in table 5). The mortality rate for algae is meant to simulate several processes that are not explicitly included, such as physiological mortality, grazing by zooplankton and other

organisms, and death due to pathogens and parasites. This rate is difficult to estimate, especially with scarce data on phytoplankton pathogens and parasites, and is a parameter that is best left for final adjustment during model calibration.

Settling of algal cells is included in the CE-QUAL-W2 model and can be positive (sinking) or negative (buoyant) to account for the ability of some algae to control their buoyancy. Settling velocities for algal particles are summarized from the literature in the CE-QUAL-W2 user manual and range from 0.0 to 30.2 m/d (Cole and Wells, 2008). Blue-green algae are unique in that many species, including AFA, are able to regulate their buoyancy with gas vacuoles and by producing and using carbohydrates and other ballast material, resulting in net settling rates that can be positive or negative during specific periods. The model algorithms in CE-QUAL-W2 do not allow the settling rates to vary over time. Algal settling rates for these Klamath River models were largely set through the calibration process.

Nutrient-limited algal growth due to lack of nitrogen or phosphorus is simulated in CE-QUAL-W2 according to a Michaelis-Menten model of half saturation. The model half-saturation coefficient for nitrogen-limited algal growth, AHSN, is defined as the inorganic nitrogen concentration (ammonium plus nitrite plus nitrate) at which the light-saturated growth rate is half of its maximum. Because certain blue-green algae, including AFA, can fix nitrogen from dissolved N_2 gas (from the atmosphere), their half-saturation coefficient for nitrogen-limited growth was set to 0, indicating that no nitrogen limitation was imposed. Nitrogen isotope $\delta^{15}N$ values for AFA collected at Link Dam in July 2008 were near 0 (Deas and others, 2009), which is an indication of N_2-fixation from the atmosphere (Vuorio and others, 2006). Calibrated values for diatoms and other algae were within the range of values cited by Cole and Wells (2008) and Padisak (2004). The model half-saturation coefficient for phosphorus-limited algal growth, AHSP, is defined as the phosphorus concentration at which the light-saturated growth rate is half of its maximum. Calibrated values for blue-green algae, diatoms, and other algae were within the range of values presented by Padisak (2004) and Cole and Wells (2008).

In addition to limitations on the algal growth rate caused by a lack of nutrients (nitrogen and phosphorus), the amount of available light is a critical factor affecting the algal growth rate. The model requires the user to set light saturation intensity at the maximum photosynthetic rate for each algal group. AFA is a "shade" species that grows well in lower light conditions due to its broader absorption spectrum. For example, AFA has been documented to grow faster than *Anabaena* under low light conditions (De Nobel and others, 1998).

Light Extinction Coefficients

Light extinction coefficients are used by the model to attenuate light through the water column, thus distributing heat vertically in the waterbody and adjusting the light intensity available for photosynthesis with depth. An accurate representation of light extinction is important for modeling both water temperature and algal growth. Light extinction by water and its dissolved constituents, by inorganic suspended solids, by organic suspended solids, and by each group of algae all contribute to the total light extinction coefficient, K_{tot}:

$$K_{tot} = K_{H2O} + K_{ISS} + K_{POM} + K_{BG} + K_{DT} + K_{oth}, \qquad (3)$$

where the contributions to the total extinction coefficient are indicated by the subscripts of the individual terms:

H2O is contribution due to water and dissolved constituents,

ISS is contribution due to inorganic suspended sediments,

POM is contribution due to suspended particulate organic matter,

BG is contribution due to blue-green algae,

DT is contribution due to diatoms, and

oth is contribution due to other algae.

The model requires separate light extinction coefficients for each contributing factor, thus computing the total light extinction coefficient as a function of the amount of each substance present:

$$K_{tot} = K_{H2O} + \varepsilon_{ISS}[ISS] + \varepsilon_{POM}[POM] + \varepsilon_{BG}[BG] + \varepsilon_{DT}[DT] + \varepsilon_{oth}[oth], \qquad (4)$$

where ε is the extinction coefficient for the component and the value in square brackets [ ] refers to the concentration of the component.

To derive the component light extinction coefficients (ε) for model input, the total light extinction coefficient K_{tot} was first determined for the existing data during April–November of 2007 and 2008 at the Railroad Bridge, Miller Island, KRS12a, and Keno sites. K_{tot} can be determined from light intensity measurements collected as a function of depth and an application of the Beer–Lambert law of light absorption. Alternatively, Secchi depth measurements, made by lowering a Secchi disk into the water and recording the maximum depth at which its pattern is still visible from the surface, can be used to estimate the total light extinction coefficient using equations such as those by Poole and Atkins (1929), Idso and Gilbert, (1974) or by Williams and others (1980). The light meter and Secchi disk methods were compared for the Link–Keno reach using a dataset from Deas and Vaughn (2006). The resulting light extinction coefficients calculated from light meter data were similar to light extinction coefficients derived from Secchi disk data and a modified Poole and Atkins equation (fig. 6). Secchi disk data collected by the Reclamation field crew during each water-sample collection trip were used with the Poole and Atkins equation to estimate total light extinction coefficients at Klamath River nonboundary sites.

After computing the total light extinction coefficient values, the results were correlated with the component concentrations (ISS, POM, and algae measured or estimated as described in the previous discussion) using multiple linear-regression methods to derive estimates of the component light extinction coefficients (ε) and K_{H2O}, the baseline light extinction due to water and dissolved substances (the regression intercept). The multiple linear-regression analysis was performed using the R statistical software (http://www.r-project.org/) to extract the extinction coefficients for each component (tables 4 and 5, values for EXH2O, EXSS, EXOM, and EXA). The derived extinction coefficient for water was 1.217/m, a relatively large value that is indicative of, and consistent with, a substantial concentration of dissolved organic matter that imparts a color to the upper Klamath River. The value of BETA, which describes how much solar radiation is absorbed at the water surface, was set to the default value of 0.45.

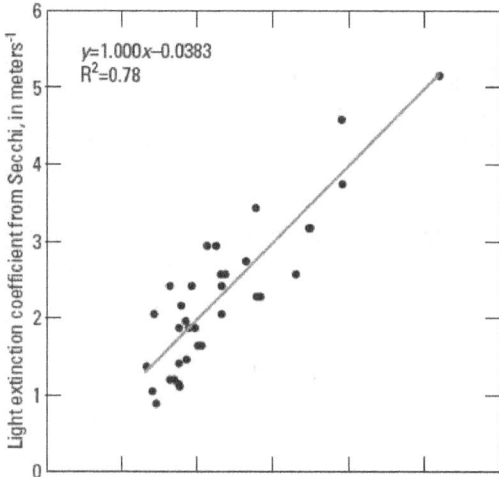

Figure 6.   Regression between light extinction coefficients derived from light meter profiles and light extinction coefficients derived from Secchi disc measurements and the modified Poole–Atkins (1929) equation for the Klamath River upstream of Keno Dam, Oregon. Light meter and Secchi disc data were obtained from Deas and Vaughn (2006).

Organic Matter Decay and Settling Rates

Decay and settling rates for particulate organic matter are required model inputs for CE-QUAL-W2 that are critically important for the Link–Keno reach of the Klamath River. Organic matter decay rates were estimated from the results of BOD experiments (Sullivan and others, 2010) that included both unfiltered and coarse-filtered samples. The Klamath River BOD samples had both labile and refractory components that were closely related to the concentrations of particulate and dissolved organic matter in those samples. Model values for several organic matter decay rates in the model (LDOMDK, LPOMDK in table 4) were derived from these experiments, with refinement during model calibration. Experiments indicated that refractory decay was slow; the values used in the model also were adjusted during calibration.

Studies were conducted in the Klamath River in 2008 to provide reach-specific settling rate data for the model (Deas and Vaughn, 2011). A Laser In-Situ Scattering and Transmissometry with Settling Tube (LISST-ST) instrument (Sequoia) was deployed in the Link–Keno reach at the Railroad Bridge sampling site. *In-situ* settling rates were assessed from three 1-week experiments that were performed during August and September 2008. Depending on the

sampling date and particle size, settling rates ranged from less than 0.2 to greater than 100 m/d for particle sizes ranging from less than 5 to greater than 350 μm. These settling rates represented all particulate matter in the water column, including dead organic matter, living algae, and zooplankton.

Larger particles (>350 μm) formed a large fraction of the particulate matter entering the Klamath River from Link River. These particles tended to have high settling rates, ranging from less than a meter per day to tens of meters per day or higher. An examination of the data from all experiments identified a median settling rate of 2.5 m/d. Smaller size fractions were examined (<62 μm, where 62 μm is the breakpoint between silt and sand) (American Society of Civil Engineers, 2000), to provide further insights into estimating settling rates for consideration in the model. The median measured settling rate was 0.44 m/d for particle sizes less than 62 μm.

Additional information gleaned during LISST-ST deployment and from more recent, detailed modeling focusing on the Lake Ewauna part of the Link–Keno reach indicates that other factors important to settling rate estimation may be occurring near or at the Railroad Bridge site. First, Link River introduces a disproportionate fraction of large particulate matter (> 350 μm), which is consistent with field data (Sullivan and others, 2008 and 2009). Second, the Lake Ewauna reach has complex hydrodynamics, including the formation of gyres, potentially influencing particle size distribution and settling rates. Finally, higher river flows affect turbulence, can resuspend some particulate matter, and thereby affect measured settling rates of particles. Reclamation conducted a surface-water spill study from July 28 through October 3, 2008, to examine methods to reduce entrainment of certain fish species (Marine and Lappe, 2009). Sub-daily flow ranges during this period ranged from less than 500 to about 3,000 ft³/s, but typically ranged from 500 to 1,500 or 2,000 ft³/s. Aside from the spill study, flows generally were stable throughout the day.

This proximity of the Railroad Bridge site to Link River can result in larger particles being present at this location than in downstream reaches of the river. This condition can be exacerbated by local hydrodynamics of Lake Ewauna proper. Furthermore, the dynamic, high-flow conditions imposed by Reclamation during the 2008 flow experiment could have reduced travel time through Lake Ewauna and re-entrained settled material from the bed. Because CE-QUAL-W2 does not accommodate variable settling rates (temporally or spatially), an average or representative settling rate was considered for the entire Link–Keno reach. The ultimate settling rate for particulate organic matter applied at the reach scale was 0.25 m/d, although actual settling rates may be higher in the upstream part of the system.

Phosphorus and Nitrogen

The CE-QUAL-W2 model includes several algorithms to mimic denitrification and the oxic or anoxic releases of nutrients in an attempt to simulate nutrient fluxes at the sediment–water interface. In one such process, the release of phosphorus from sediments under anoxic conditions (PO4R, table 4) is specified as a fraction of the model's zero-order SOD rate. Conversely, nutrients can be released to the water column under oxic conditions by the decay of organic matter in the sediments compartment, which creates the model's first-order SOD.

Loss of nitrate and inorganic nitrogen through denitrification can occur where oxygen conditions are depleted and specialized bacteria use nitrate (rather than oxygen) as a terminal electron acceptor. The model also can approximate denitrification through an optional mechanism that allows nitrate to diffuse into the sediments where it is converted or utilized. The release rate of ammonia from sediments under anoxic conditions (NH4R, table 4) is specified in the model as a fraction of the zero-order SOD rate. The oxidation rate of ammonia (nitrification) also is a specified model parameter.

Sediment Oxygen Demand

Rates of consumption of dissolved oxygen at the sediment–water interface (SOD) have been measured in the Link–Keno reach of the Klamath River by several groups of researchers. *In-situ* SOD studies were conducted in this reach in early June 2003, prior to the influx of the blue-green algae from Upper Klamath Lake (Doyle and Lynch, 2005); the median measured SOD rate was 1.8 (g O_2/m^2)/d at 20 °C. Laboratory-based SOD studies (Raymond and Eilers, 2004; Eilers and Raymond, 2005) using sediment cores gave similar values.

The CE-QUAL-W2 model has two mechanisms to simulate SOD: zero-order and first-order. The zero-order process allows the sediment to exert a relatively constant oxygen demand, with the rate modified only by water temperature. First-order SOD, however, keeps track of all organic matter and algae that settle to the sediment. More organic matter in the sediment compartment means more oxygen demand, which is important for simulating the response of a large flux of settling algae in the Klamath River in midsummer. The CE-QUAL-W2 model calculates a decay rate for the first-order sediment compartment based on the individual decay rates of algae, LPOM, and RPOM, and weighted by the amount of each that has settled to the river bottom. Both SOD mechanisms, but especially the first-order demand, were used in these models to produce a dynamic representation of the seasonal variation in SOD. The SOD temperature factors also were adjusted during model calibration.

Reaeration

The model user must select one of many possible equations in CE-QUAL-W2 to represent the reaeration of oxygen across the air–water interface. Rivers are usually assigned reaeration equations based on river depth and water velocity. Lakes are usually assigned reaeration equations based on depth and wind speed. A number of reaeration equation options were tested; the lake equation 14 was used in the final model, due to the low velocities and strong effect of wind in this reach:

$$K_a = \frac{c_1 + c_2 w^{c_3}}{H},\qquad(5)$$

where
  w is the wind speed (meters per second),
  H is the channel depth (meters),
    and the constants in this equation
    (c_1, c_2, c_3, table 4) were optimized with use of the PEST parameter optimization software (Doherty, 2010).

Model Results

During model calibration, a subset of model parameters was adjusted within reasonable bounds to optimize a comparison of measured and modeled results for this river reach. Manual model calibration consisted of iterative model runs where the values of key model parameters were adjusted to better fit the results during unique time periods where one or more processes could be somewhat isolated from the many processes affecting water quality throughout the year. The magnitude and seasonal and spatial patterns between measured data and model output should correspond; however, an exact match is not expected because measurements are made at a point and model output represents conditions in model cells that encompass about 1,000 ft of river length. Concentration values for suspended particulate material, including algae, particulate carbon and nitrogen, and total nitrogen and phosphorus, had more spatial variability than values for dissolved constituents (Sullivan and others, 2008, 2009); the model is laterally averaged and does not simulate any bank-to-bank differences. Models are simplifications of nature, and not every process is simulated, but a good fit between the spatial and temporal patterns in the modeled and measured data provides confidence that major processes are included and that the model can be used to provide insight into system behavior under various management options.

Model calibration began with balancing the water budget followed by assessing water velocity, ice cover, and water temperature. Finally, the water-quality aspects of the model were calibrated, beginning with some of the constituents that exhibited conservative (nonreactive) properties, such as TDS and ISS, then moving on to nutrients (nitrate, ammonia, dissolved phosphorus) and organic matter, and finally addressing the more challenging constituents such as algae and dissolved oxygen. pH was initially included in the model, but is not considered to be calibrated. This system has high concentrations of dissolved organic matter which can affect pH and buffering capacity, as indicated theoretically and also from the results of alkalinity titrations in this reach (U.S. Geological Survey, Miller Island, unpub. data, 2007). The CE-QUAL-W2 model does not yet include algorithms that include the acid–base properties of organic matter in the simulation of pH.

The parameter optimization software PEST (Doherty, 2010) was linked to CE-QUAL-W2 during the model calibration process and thus was used to refine the model calibration and to examine parameter sensitivity and parameter correlation. Much knowledge was obtained about the nature and characteristics of water-quality processes acting in the Klamath River during the model calibration process.

Water Velocity

Simulation of water velocity allows the calculation of travel time through the Link–Keno reach. Slower velocities allow more time for settling, decay, and nutrient transformation to occur within the reach; faster velocities move more material downstream. Generally, the variation in velocity from bank to bank was low and less than the vertical variability in velocity for most of the Lake Ewauna to Keno Dam reach according to the 2007 cross-sectional measurements. Flow direction usually was aligned with the upstream–downstream axis of the channel. The lack of substantial lateral variability and the alignment of velocities with the longitudinal axis of the river confirmed that the two-dimensional model CE-QUAL-W2 was a suitable choice for the Link–Keno reach. Although average measured cross-sectional velocity was not directly comparable to simulated velocity, because model velocity is a representation of average conditions that occur over about 1,000 ft of river length, measured and modeled velocities were in the same range (fig. 7). Modeled and cross-sectional average measured velocities ranged from near 0 to 0.5 ft/s on the measurement dates in 2007. September velocities were lower than May velocities, which was expected because streamflow in this reach tends to be lowest in August and September. Maximum modeled velocity (volume averaged over segment depth) in 2006–09 for the whole modeled Link–Keno reach was 3.0 ft/s near the Keno sampling site in mid-April 2006 during high flow; minimum modeled velocity was near 0 ft/s at the upper

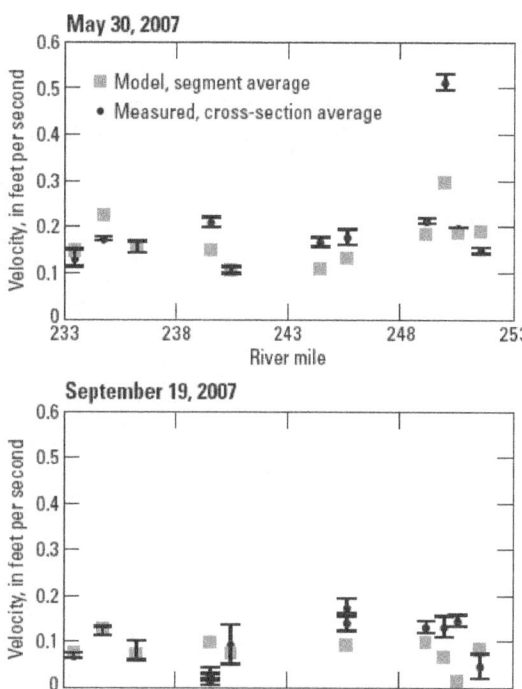

Figure 7. Measured and modeled average cross-sectional velocity at sites in the Link River to Keno Dam reach of the Klamath River, Oregon, May 30 and September 19, 2007. Cross-sectional velocities were measured at specific locations, but model output is an average velocity of a model segment. Segment length varies, but most encompass around 1,000 feet of the river.

end of Lake Ewauna at several times over the modeled period. Generally, modeled velocities were highest near the Lost River Diversion Channel inflow and near the Keno sampling site, and lowest in Lake Ewauna and midreach near the Miller Island and KRS12a sites. Simulated travel times from model tracer tests for the Link River to Keno Dam reach ranged from about 4 days at 2,000 ft³/s flow to 12 days at 700 ft³/s flow.

The 30-minute ADCP measurements at the Keno site also showed that velocity direction was aligned with the longitudinal axis of the channel (fig. 8A). Although measured and modeled velocity are not directly comparable, results in figure 8B show that the two are similar in magnitude and that the temporal patterns in the measured data are well represented by the model. Daily average modeled and measured velocities from June to early December 2008 ranged from near 0 to almost 0.6 ft/s. Velocity variations at the Keno site over periods of hours to days were the result of changes in operations at Keno Dam; abrupt changes in flow at the dam led to abrupt changes in upstream water velocity.

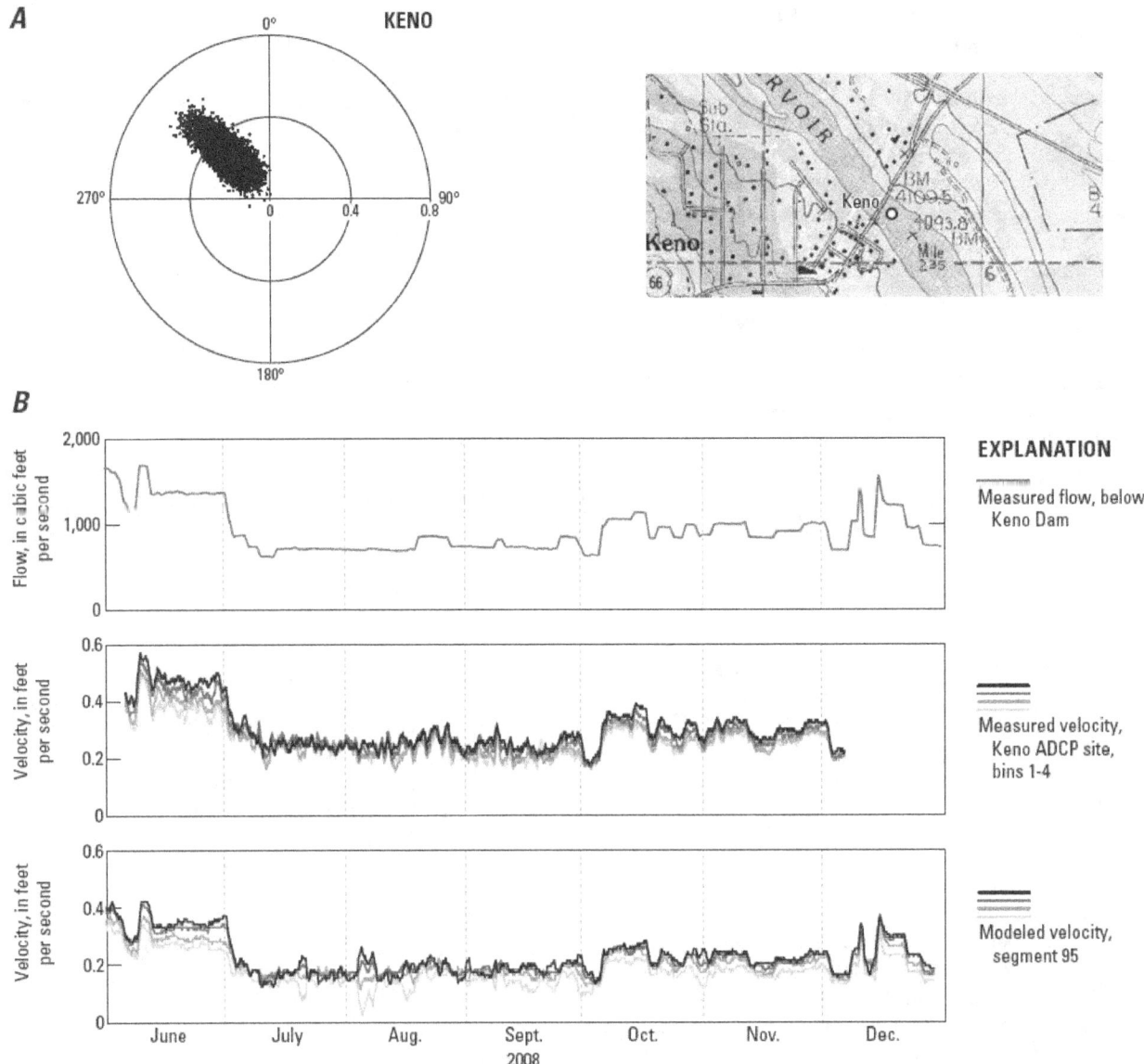

Figure 8.   (*A*) Measured flow direction (degrees) and magnitude (feet per second) of acoustic Doppler current profiler (ADCP) velocity measurements at the Keno site, Oregon, early June through early December 2008. (*B*) Daily moving average of modeled and measured velocity at the Keno site. Measurements are at a specific location; model output is average velocity of a model segment encompassing 1,030 feet of river length. Darker lines represent velocity nearer the surface, paler lines represent velocities nearer the river bottom.

The hydrodynamics of the Lake Ewauna reach were complex. Flow often was aligned with the upstream–downstream axis of the river channel, but the direction of flow at EWA1 and EWA2 periodically would reverse (fig. 9A). This flow pattern was affected by both the inflow from Link River, just upstream at the northwest corner of figure 9A, and by the wind direction. For June and July 2008, with relatively higher flows from Link River, when the wind direction was towards the southeast, in the same direction as the flow from Link River, flow at both EWA1 and EWA2 usually was in the downstream direction (fig. 9B). If the wind was towards the northwest for the same period, flow at the deeper western

side (EWA1) was still towards the southeast, but flow on the shallow eastern side (EWA2) was reversed towards the northwest, thus suggesting a counterclockwise circulation pattern. During mid-July through October, with lower flow from Link River, periods still occurred when flow at both EWA1 and EWA2 were in the downstream direction, but wind had a greater influence on circulation. For example, sometimes when wind direction was towards the northwest, a strong counterclockwise circulation was suggested, with flow at EWA1 to the southeast and flow at EWA2 to the northwest. At other times, when wind direction was towards the southeast, flow at EWA1 was to the northwest and flow at EWA2 to

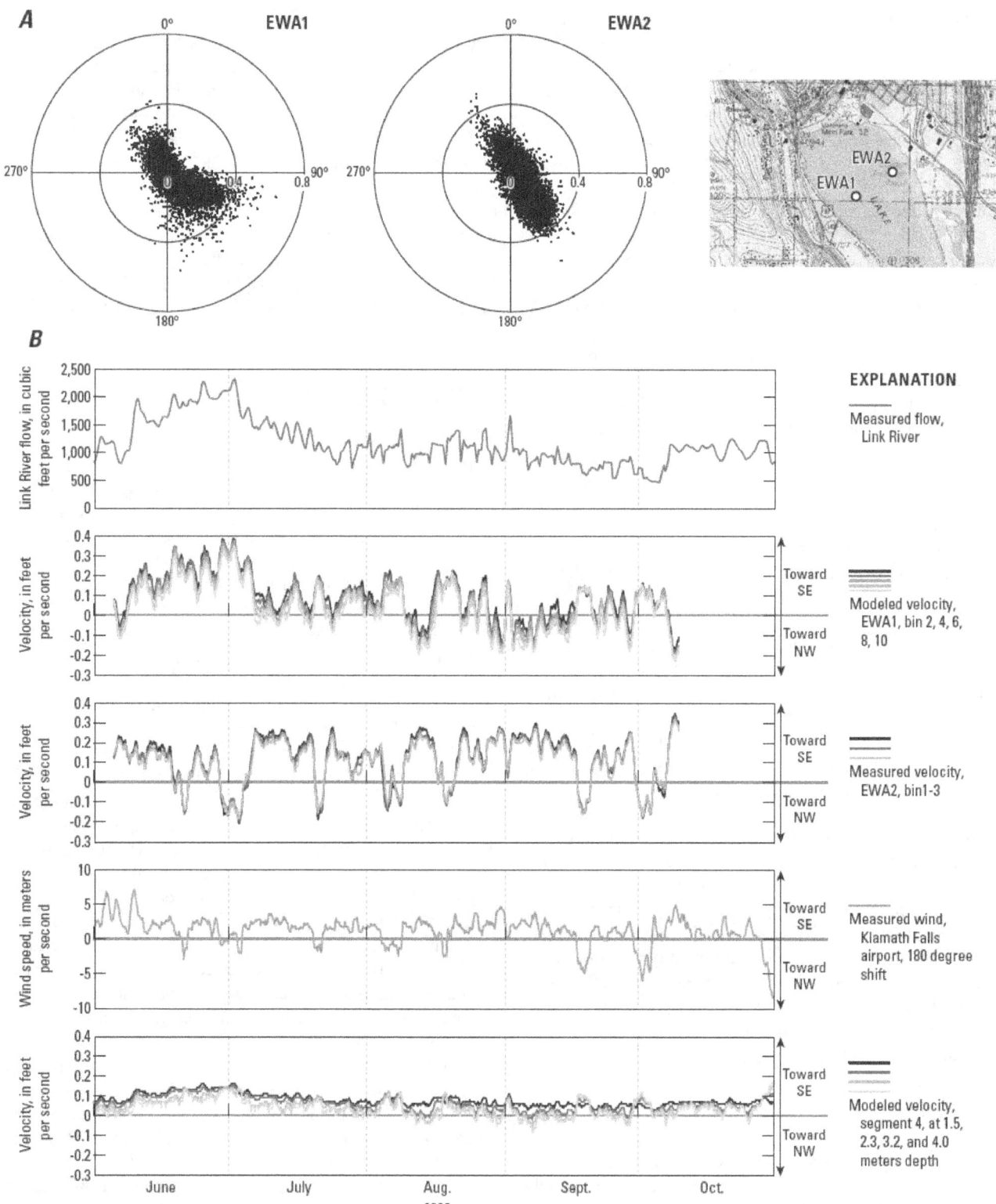

Figure 9. (*A*) Measured direction (degrees) and magnitude (feet per second) of ADCP velocity measurements at the Ewauna sites, Oregon, for early June through October 2008. (*B*) Measured flow at Link River, Oregon, daily moving average measured velocities at EWA1 and EWA2 in Lake Ewauna, Oregon (negative flow is to the northwest, positive to the southeast), daily average measured wind (shifted 180 degrees so that, similar to flow, wind to the northwest is negative), and modeled velocity. Measured velocities are at specific locations; model output is average velocity of a model segment encompassing 1,134 feet of river length. Darker lines represent velocity nearer the surface, paler lines nearer the river bottom.

the southeast, thus suggesting clockwise circulation. Flow at EWA2 tended to be in the same direction as the wind, whereas flow at EWA1 was downstream during high flow, but could be reversed at low flow when winds were toward the southeast.

For a shallow lake with a trench to the west, clockwise circulation with wind towards the southeast, and counterclockwise circulation with wind towards the northwest is expected theoretically (Ji and Jin, 2006). When most wind is toward the southeast, a similar clockwise pattern dominates the circulation in Upper Klamath Lake (Wood and others, 2008), a shallow lake that also has a deeper trench on its western edge. The circulation pattern in Lake Ewauna, like that in Upper Klamath Lake, was strongly affected by the wind, but was complicated by the strong advective flow from Link River, especially in the spring. The two ADCPs have provided valuable insight into the complex circulation in Lake Ewauna. Additional ADCP measurements could be valuable to gain a more detailed understanding of these circulation patterns.

The two-dimensional CE-QUAL-W2 model cannot capture the details of these flow patterns in Lake Ewauna because lateral variations are not simulated. Instead of simulating higher flows varying between upstream and downstream direction, the model simulates an average flow from bank-to-bank, most often in the downstream direction. The complex flow patterns in Lake Ewauna result in a wide distribution of residence times there, with some water parcels that traverse quickly downstream and bypass much of the large volume of the reach and other water that becomes entrained in a recirculating flow and spends a longer time there. An increased or decreased travel time has an important effect on water quality because more or less time is available for settling or decay or other processes. The model, in contrast, will simulate a narrower distribution of travel times and, therefore, a narrower distribution of water-quality effects, although the average effect on water quality may not be biased. Although the circulation pattern in Lake Ewauna undoubtedly has some interesting water-quality effects, it was decided that because the reach is relatively short, the increased effort and complexity associated with using a different type of model for the Lake Ewauna reach was not appropriate for this study. A more detailed examination of water-quality processes in Lake Ewauna in the future could benefit from the development of a three-dimensional model.

Ice Cover

Ice forms at times on the Link–Keno reach of the Klamath River, because winter air temperatures are commonly below freezing. The CE-QUAL-W2 model can simulate the onset, buildup, and break up of ice. These processes in the model are affected by the locations and temperatures of inflows and outflows, the ice-to-water heat-exchange coefficient, and sublimation effects of wind over the ice surface. Proper simulation of ice cover is important to the accurate simulation of water temperature and winter reaeration.

The model simulated ice to have occurred on some segments and dates in December 2006, January–February and December 2007, January–February and December 2008, and January and December 2009. This simulation matches well with field observations of the presence of ice by the field crews. At times, the crew noted partial ice cover along the banks. Because CE-QUAL-W2 is a laterally averaged model, simulation of lateral differences from bank-to-bank were not represented. The model would instead simulate partial lateral ice cover as a thin layer of continuous ice from bank to bank. The model does simulate differences in ice-cover from upstream to downstream and those spatial and temporal patterns of ice cover simulated by the model match well with the available field observations.

Water Temperature

Water temperature in rivers is affected by the temperature and flow of inflows and outflows, heat exchange at the air–water interface, light extinction, and mixing by wind. Water temperature is usually one of the first constituents to be calibrated in CE-QUAL-W2 model applications because many other simulated water-quality processes, such as chemical reactions, algal growth, and SOD, are dependent on temperature. The physics of heat-exchange processes for waterbodies is well known and has been accurately translated into mathematical formulas and algorithms that are readily coded into numerical models. The CE-QUAL-W2 model includes the most important algorithms describing heat-transport and exchange processes.

The seasonal and daily temperature cycles in the Link–Keno reach were consistent in the 4 years simulated for this study (figs. 10 and 11), with cold winter water temperatures near freezing and warm summer water temperatures near 30 °C at times. Seasonal water temperature patterns were largely a result of the inflow from Upper Klamath Lake, the large shallow lake just upstream; seasonal temperature patterns were similar at the upstream and downstream ends of the reach.

Daily temperature cycles were present in summer, and heating of the river surface is enhanced in the Klamath River because of high light extinction coefficients that cause most of the short-wave solar radiation to be scattered, absorbed, and converted to heat near the top of the water column. For most of the year and at most locations in the Link–Keno reach, temperature differences between the river top and bottom were present only during daytime; at night, cooling of the river surface often decreased vertical temperature differences sufficiently to allow vertical mixing. At some locations, especially in the downstream reaches, temperature differences of up to several degrees Celsius between the river surface and bottom persisted, up to about 1–2 weeks, thus inhibiting vertical mixing. Intermittent thermal stratification downstream of the Klamath Straits Drain inflow may be enhanced by the sinking of that inflow due to its higher TDS concentration (discussed in the next section, "Total Dissolved Solids and Specific Conductance").

A   2006

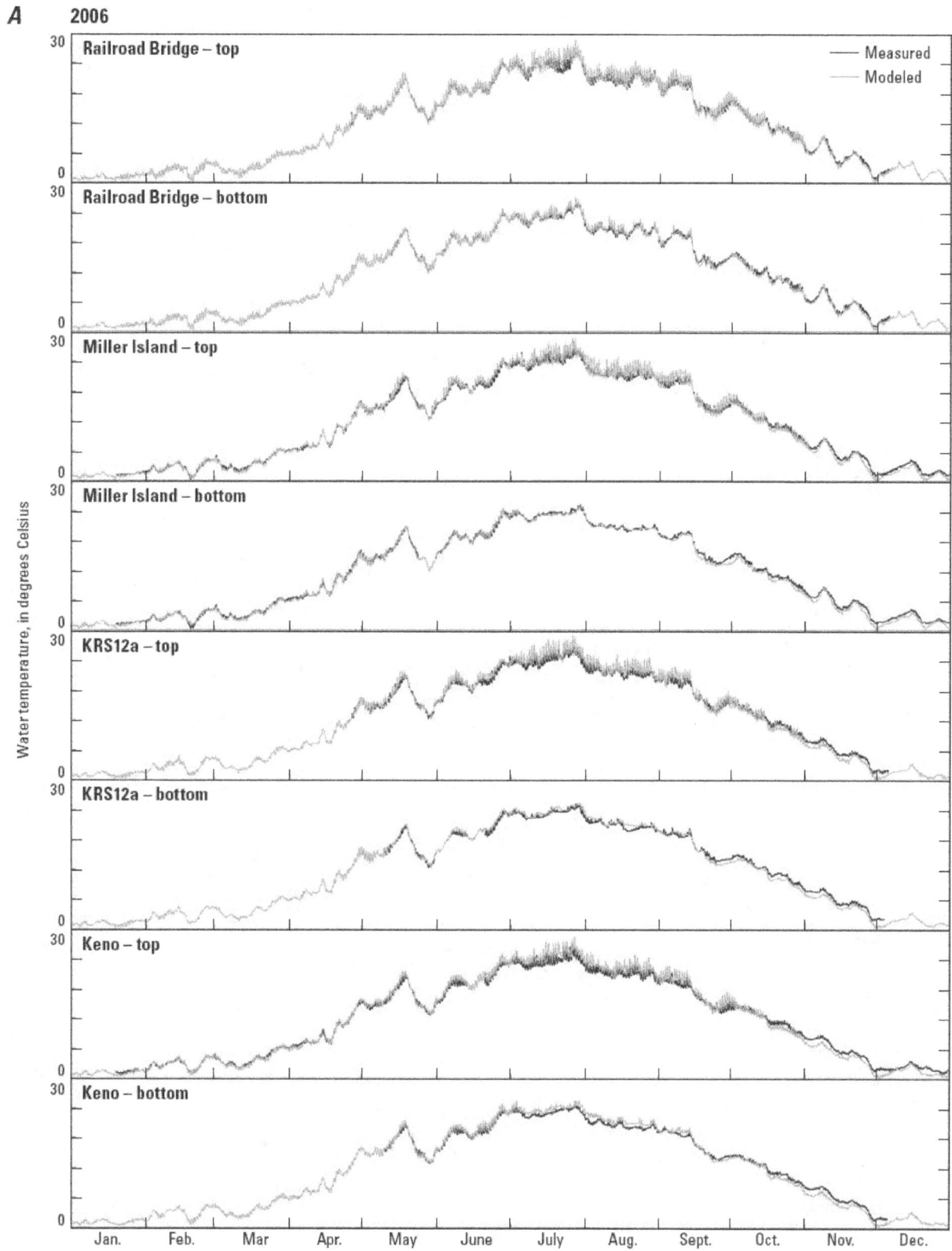

Figure 10.   Measured and modeled hourly water temperature for calendar year (*A*) 2006, (*B*) 2007, (*C*) 2008, and (*D*) 2009 for sites in the Klamath River upstream of Keno Dam, Oregon.

B   2007

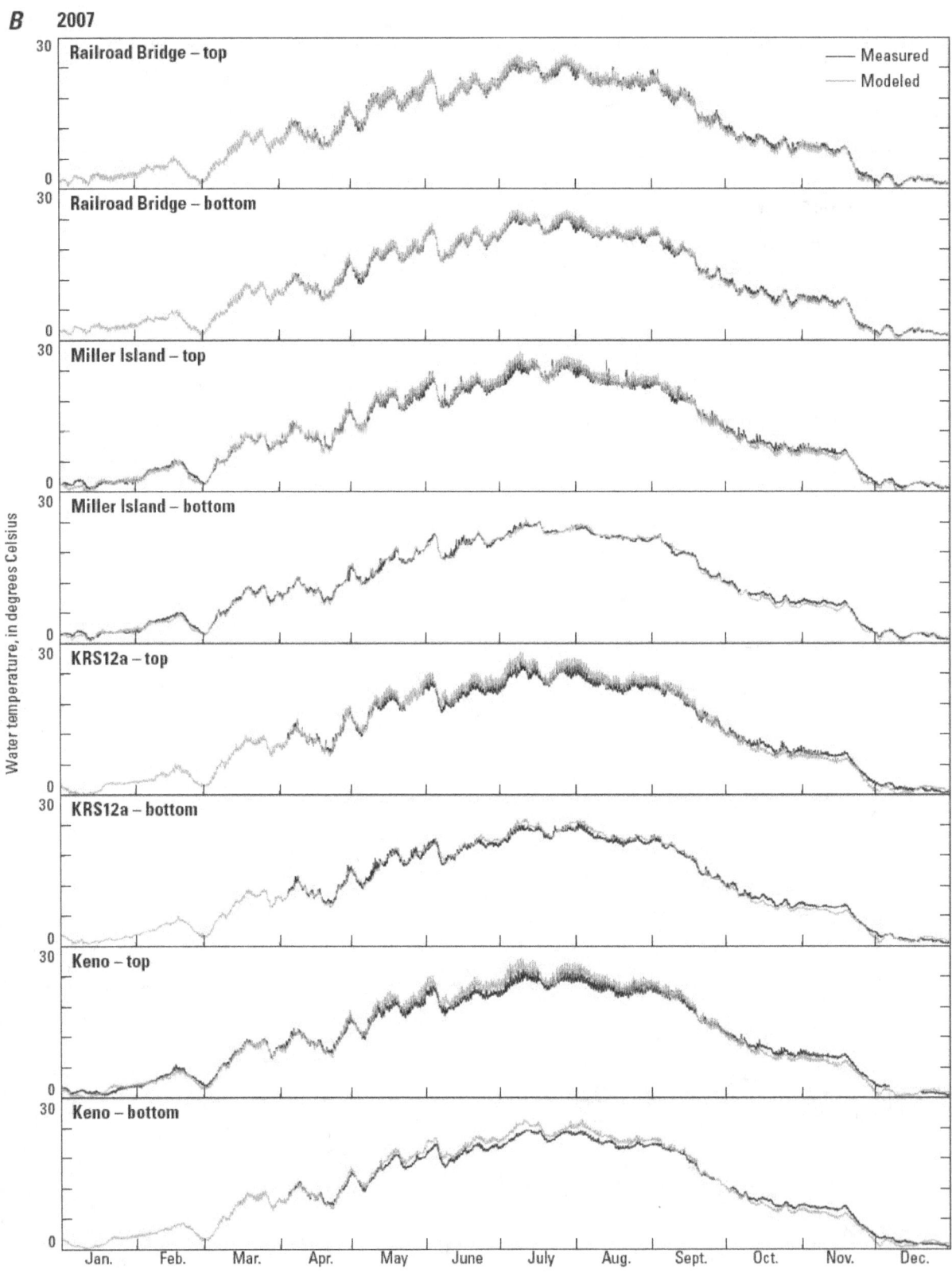

Figure 10.—Continued.

C   2008

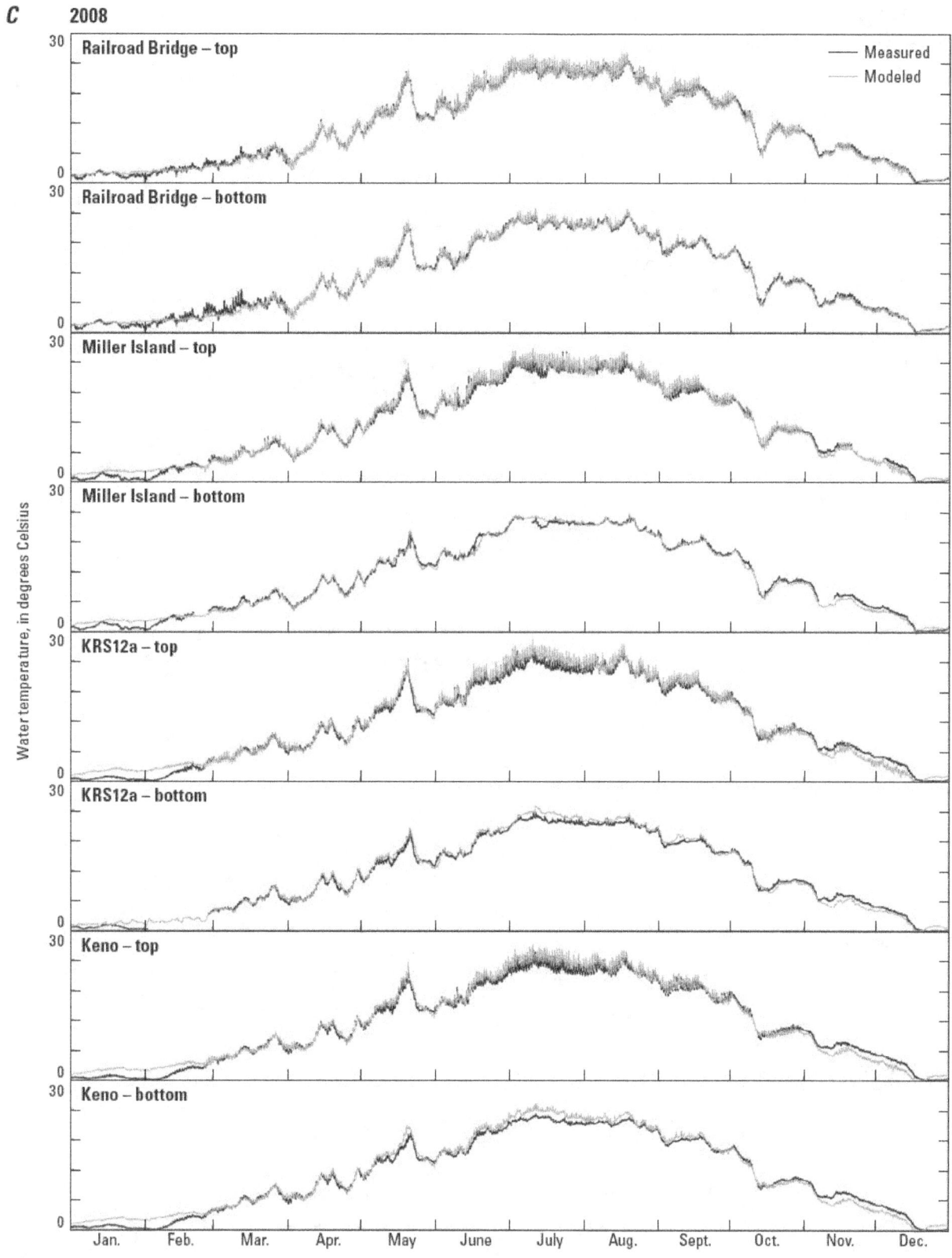

Figure 10.—Continued.

D   2009

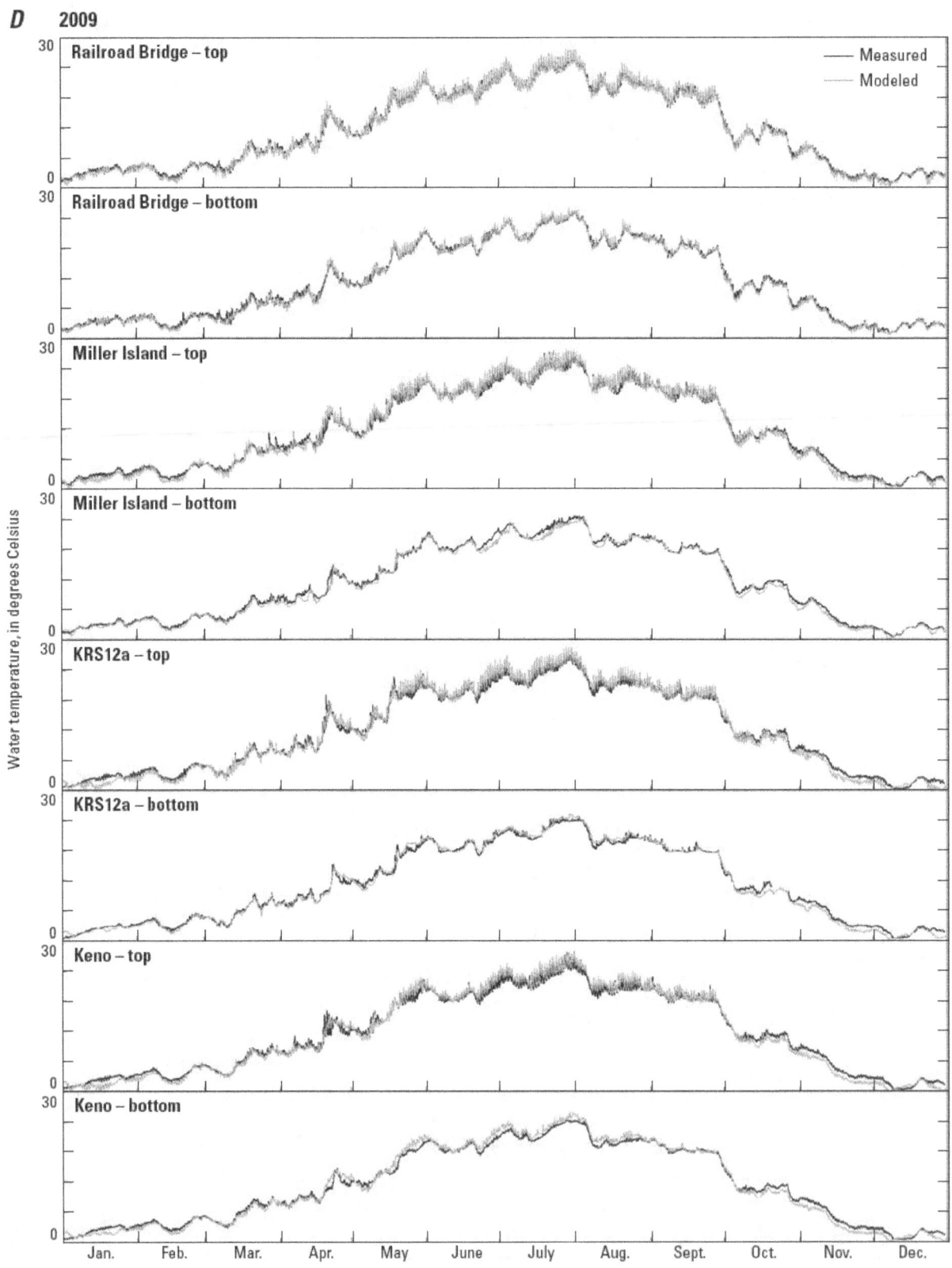

Figure 10.—Continued.

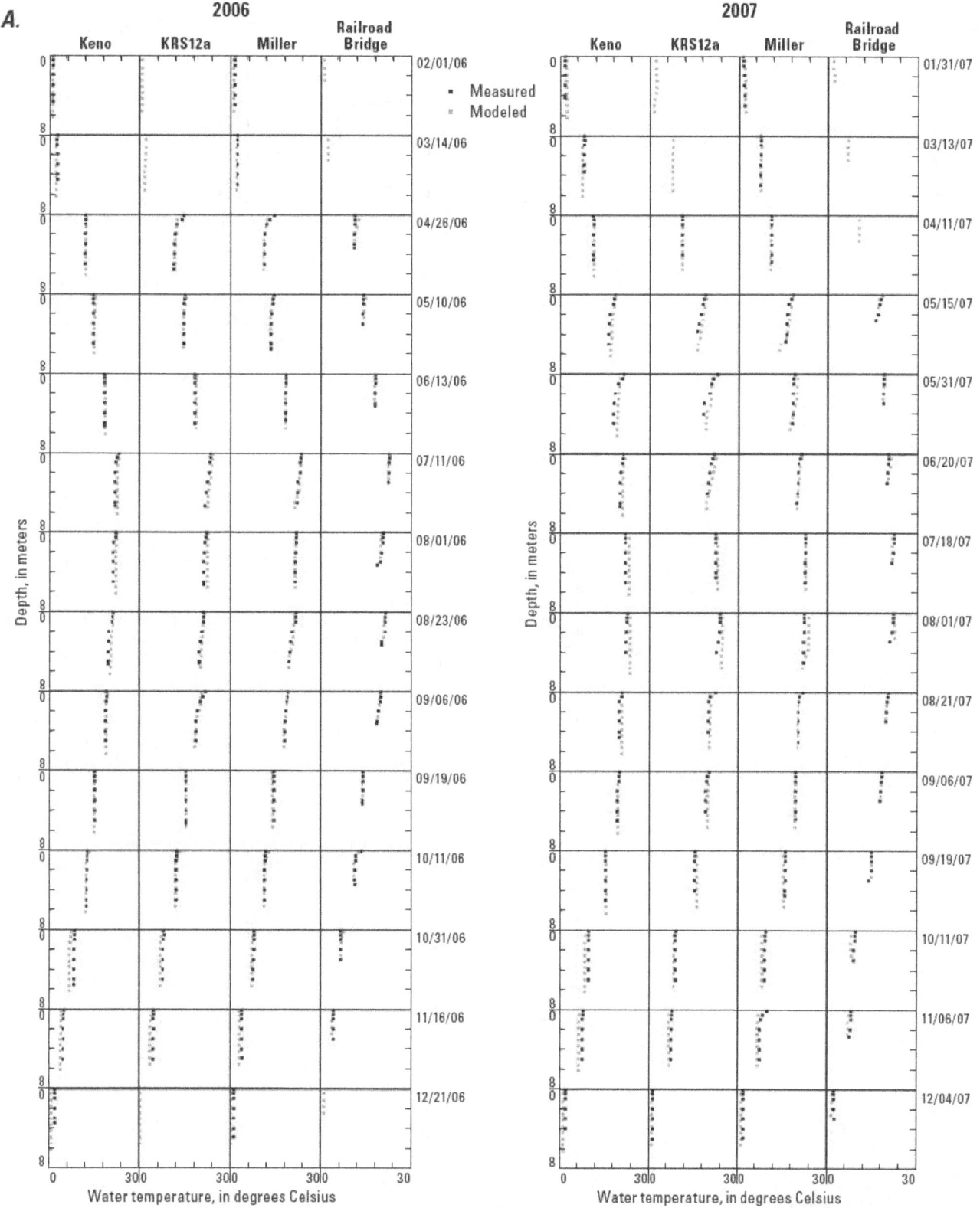

Figure 11. Profiles of measured and modeled water temperature for selected dates and sites in (*A*) 2006 and 2007, and (*B*) 2008 and 2009 in the Klamath River upstream of Keno Dam, Oregon.

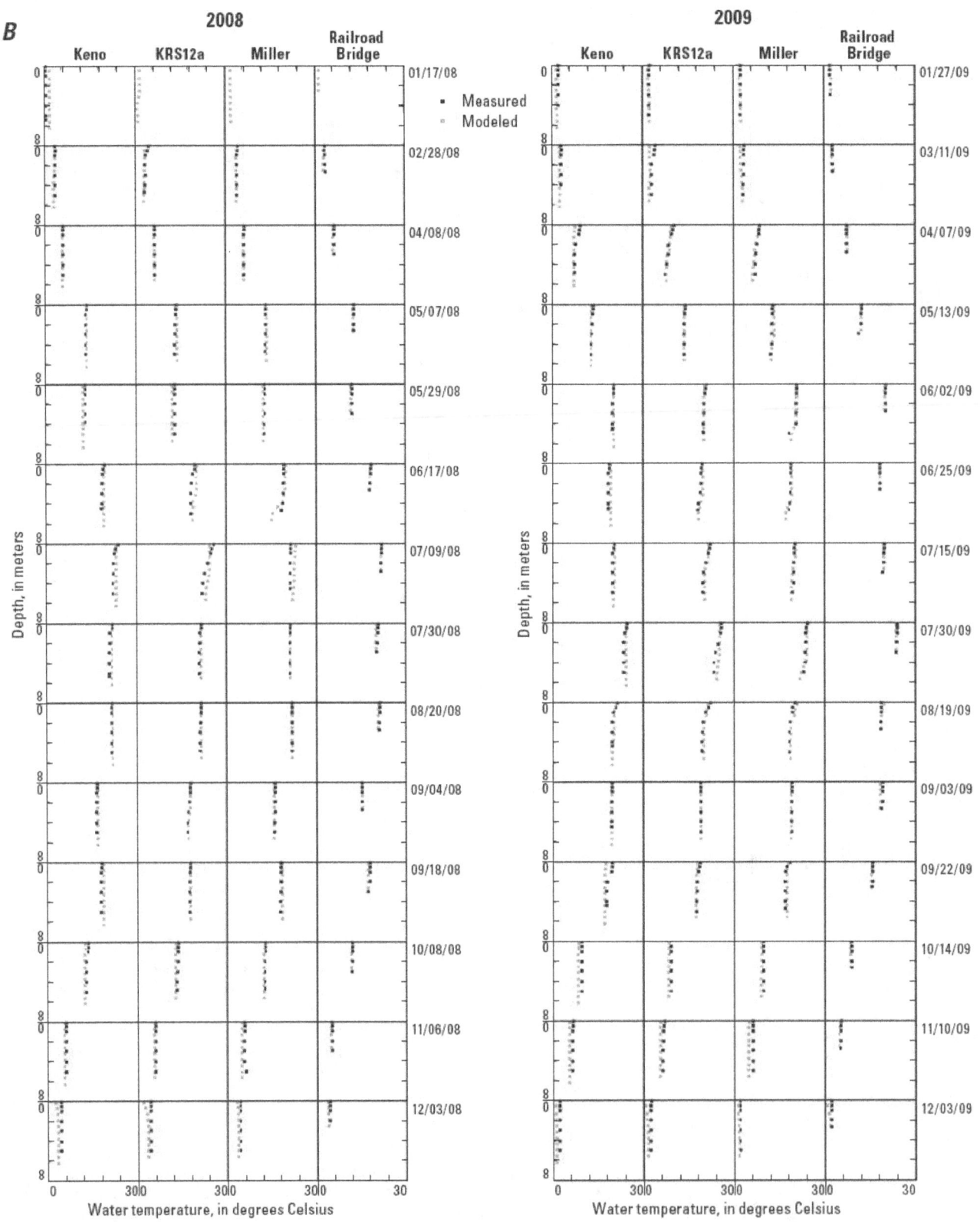

Figure 11.—Continued.

Goodness-of-fit statistics were calculated to compare hourly measured temperatures and model output at the same location and time (table 6). The mean error (ME) for 2006–09 averaged over all sites for a year was between -0.17 and 0.17 °C depending on the year. The ME is an indication of the overall bias in model predictions. An ME <0.2 °C indicates that overall bias of the model is on the same order of magnitude as measurement error, which typically is ±0.2 °C. The computed mean absolute error (MAE) for model temperature predictions was between 0.54 and 0.64 °C. The MAE is an indication of the typical error associated with any one model-data comparison. Sources of error in modeling water temperature include the estimation of Lost River Diversion Channel temperatures in 2006, estimation of distributed tributary temperatures, and any errors in the model representation of channel width or calculation of extinction coefficients. The model consistently produces a slight underprediction of water temperature in the fall, which may be due to the lack of an algorithm in CE-QUAL-W2 to transfer stored summer heat from river sediments to the water column. Overall, the model accurately simulated water temperature in all 4 years with an MAE of 0.64 °C or less; an MAE of less than 1.0 °C is the usual benchmark for a well-calibrated CE-QUAL-W2 model.

Table 6. Range of measured values at calibration sites and goodness-of-fit statistics averaged over all calibration sites.

[**Abbreviations:** C, degrees Celsius; L, liter; m, meter; mg, milligram; mL, milliliter; μm, micrometer; <, less than; rl, reporting level]

Constituent	Unit	Data type	Range of measured values	Year	Mean error	Mean absolute error	Root mean square error
Water temperature	°C	Hourly	0–28	2006	-0.01	0.54	0.69
				2007	0.06	0.64	0.78
				2008	0.17	0.63	0.80
				2009	-0.17	0.58	0.72
Dissolved oxygen	mg/L	Hourly	0–20	2006	0.12	1.12	1.57
				2007	0.51	1.30	1.84
				2008	-0.19	1.01	1.47
				2009	-0.66	1.30	1.77
Ammonia	mg/L	Grab	<rl–1.7	2007	0.05	0.12	0.17
				2008	-0.08	0.21	0.25
Nitrate	mg/L	Grab	<rl–0.8	2007	0.01	0.03	0.03
				2008	0.01	0.03	0.03
Particulate nitrogen	mg/L	Grab	0.06–4.0	2007	-0.18	0.28	0.48
				2008	-0.15	0.24	0.38
Total nitrogen	mg/L	Grab	0.6–5.9	2007	0.11	0.41	0.58
				2008	-0.03	0.38	0.48
Orthophosphorus	mg/L	Grab	0.01–0.27	2007	0.01	0.03	0.05
				2008	0.00	0.03	0.03
Total phosphorus	mg/L	Grab	0.06–0.50	2007	0.00	0.05	0.07
				2008	0.00	0.04	0.05
Blue-green algae	μm^3/mL	Grab	0–113×10^6	2007	-3.3×10^6	7.5×10^6	14.9×10^6
				2008	0.5×10^6	3.6×10^6	5.8×10^6
Particulate carbon	mg/L	Grab	0.5–18	2007	0.15	1.44	2.26
				2008	0.61	1.38	1.96
Dissolved organic carbon	mg/L	Grab	5–14	2007	0.36	0.75	1.15
				2008	-0.19	0.65	0.81

Total Dissolved Solids and Specific Conductance

Total dissolved solids (TDS) is the sum of all dissolved substances in water, such as dissolved nutrients, dissolved organic matter, and all dissolved ions. Total dissolved solids contribute to density gradients in the model, which can affect how constituents are mixed through the water column. The CE-QUAL-W2 model treats TDS as conservative (unreactive), in that its concentration is only affected by inflows, outflows and hydrodynamic mixing. This is a good approximation, but it is important to realize that TDS in rivers also is affected by a number of chemical and biological processes, and those processes do not affect TDS in the model. Total dissolved solids, as discussed in section, "Methods," are related to specific conductance; specific conductance was converted to TDS for use in the model and model output was converted back to specific conductance to compare with field calibration data.

The inflow specific conductance at Link River generally was lower than specific conductance from point sources and tributaries. Model options were set so that tributary inflows were placed in the river at a depth based on the density of the inflow and that of the river, where the densities were determined from TDS and water temperature. The effect of this density placement of inflows was most obvious at the KRS12a site downstream of the inflow of Klamath Straits Drain. The higher density Klamath Straits Drain water plunges to the bottom of the Klamath River at certain times of year; the Klamath Straits Drain has especially high specific conductance in spring. That density stratification was especially evident in the hourly data and specific conductance profiles for early 2007–09 (figs. 12 and 13). The density stratification decreases by the time the water reaches the Keno site, indicating that additional vertical mixing has occurred. Specific conductance from the NPDES point sources was as high as or higher than that from Klamath Straits Drain, but the point source flows were low enough that the density effect was not evident in the profiles at Railroad Bridge. The higher specific conductance from the NPDES point sources, however, does affect the depth at which those flows entered the river.

The model does well at simulating specific conductance at most sites (figs. 12 and 13), which indicates that inflows, transport, and mixing were the major controlling processes. Some of the small-scale daily variation in specific conductance that occurs at the Railroad Bridge site, especially Railroad Bridge-bottom, is not simulated by the model. This can indicate that the processes that produce those variations originate from a process not represented in the model.

Inorganic Suspended Sediment

Inorganic suspended sediment (ISS), such as suspended clay or silt, is a natural component of lakes and rivers, though excessive concentrations can impair certain uses of a river. For example, high ISS loads can reduce the efficiency of water treatment plants, and, over time, high concentrations of sediment can settle and impair fish habitat or reduce available water storage in a reservoir. High concentrations of ISS also affect light penetration, which affects the vertical distribution of heat and the depths at which photosynthesis can occur. Some of the tributaries in the Link–Keno reach, such as Klamath Straits Drain, had relatively high ISS concentrations over periods of several days to weeks in winter. These high concentrations usually were associated with storms and could come from stormwater picking up sediment as it runs over soil or increased flow and turbulence causing resuspension of bed sediment. In-river modeled ISS concentrations downstream of Klamath Straits Drain were elevated (above 30 mg/L) for periods in late February to early April in 2006, 2007, 2008, and 2009. ISS also was elevated for most of the entire Link–Keno reach in early December 2007 and late March to early April 2008. Summer concentrations of suspended sediment typically were low in this reach at about 3–5 mg/L.

A   **2006**

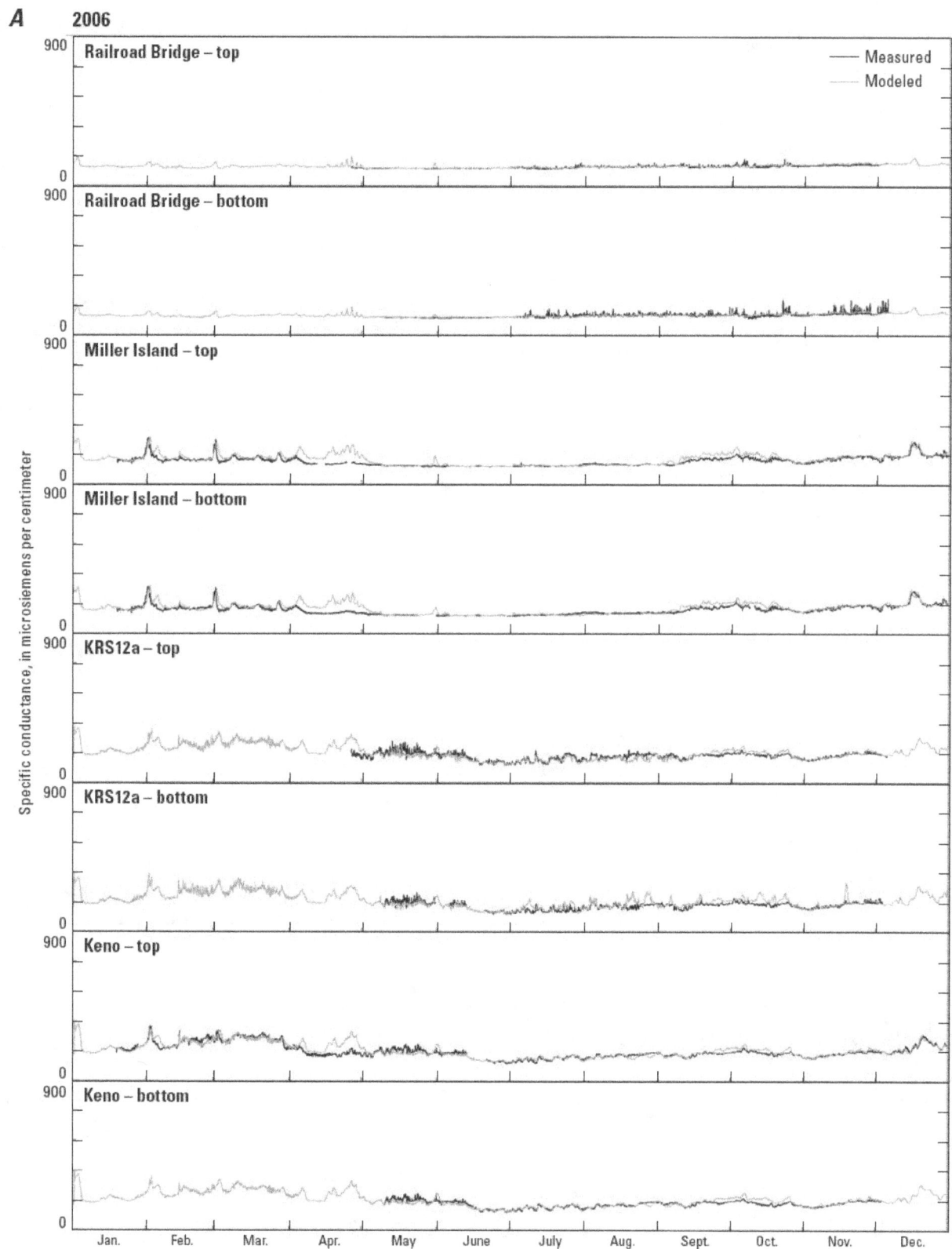

Figure 12. Measured and modeled hourly specific conductance for calendar years (*A*) 2006, (*B*) 2007, (*C*) 2008, and (*D*) 2009 for sites in the Klamath River upstream of Keno Dam, Oregon.

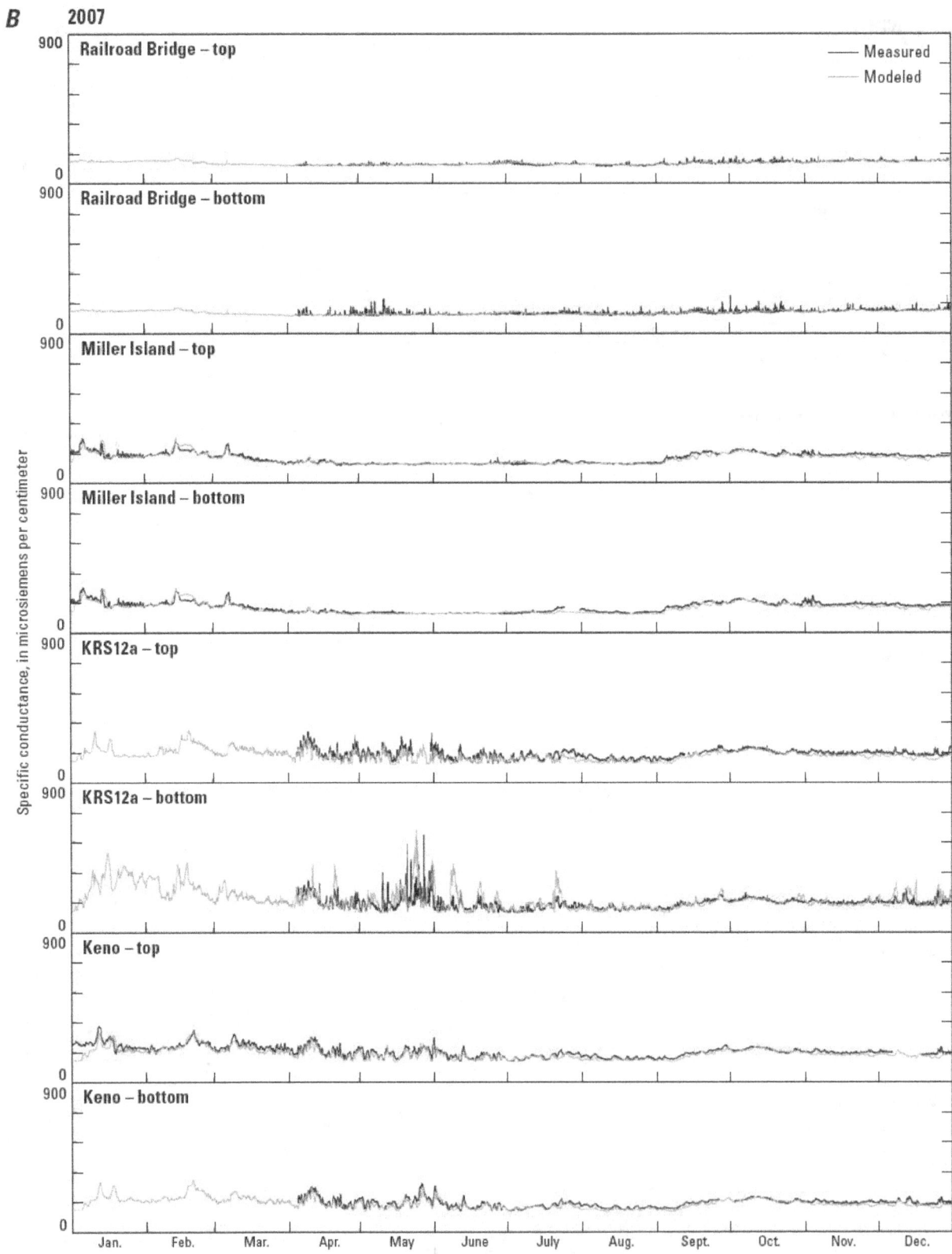

Figure 12.—Continued.

C   2008

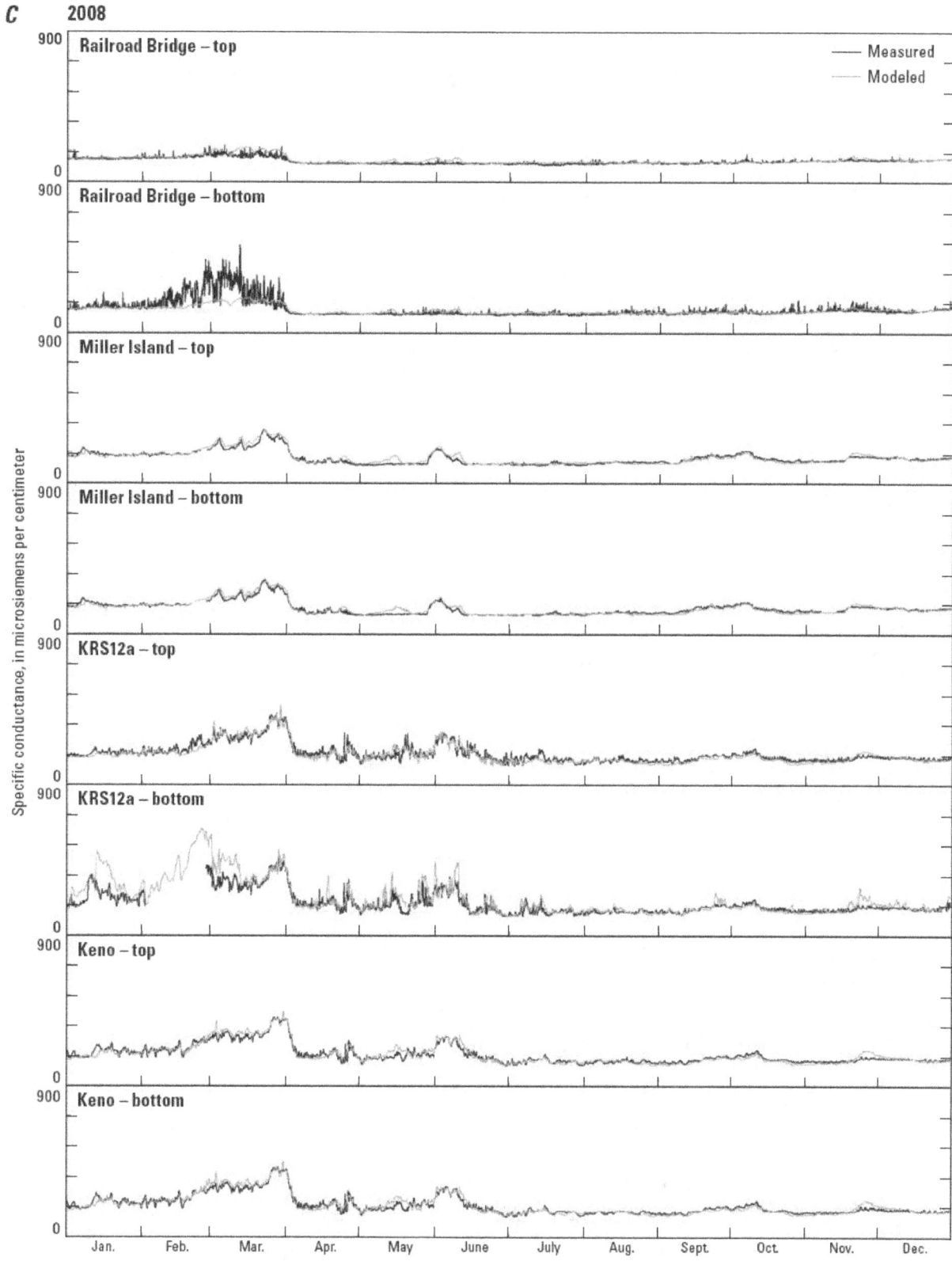

Figure 12.—Continued.

D   2009

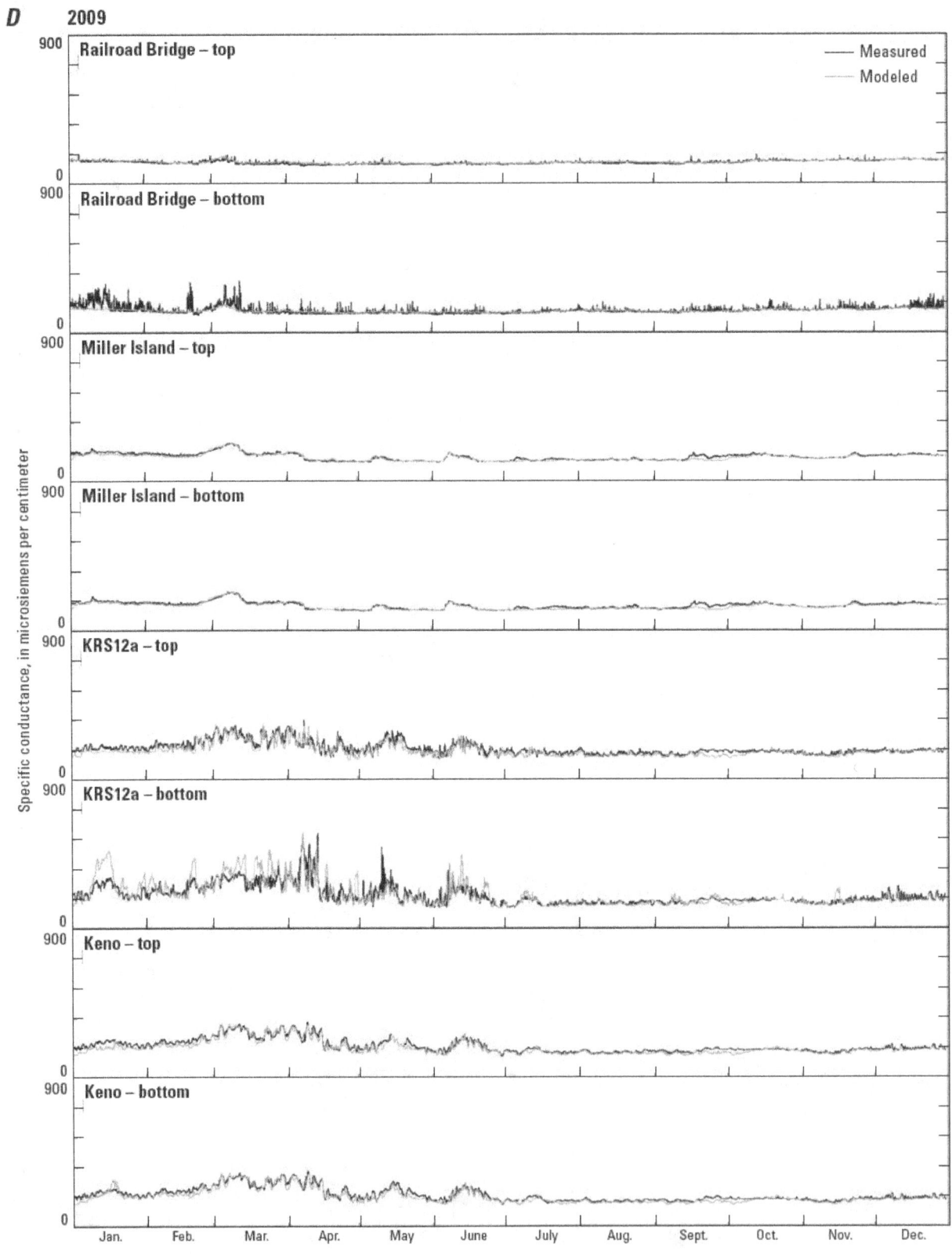

Figure 12.—Continued.

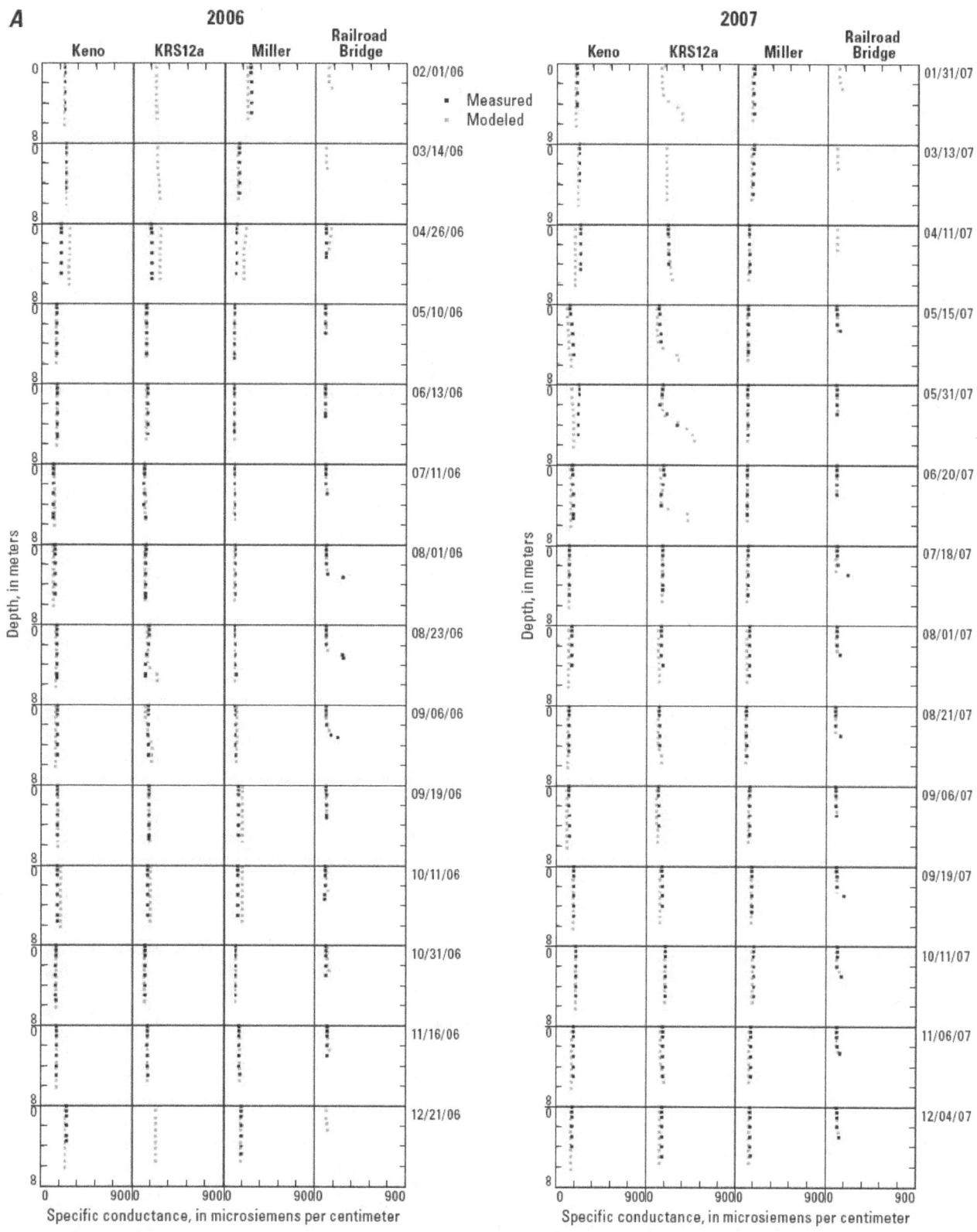

Figure 13. Measured and modeled specific conductance for selected dates and sites in (*A*) 2006 and 2007, and (*B*) 2008 and 2009 for the Klamath River upstream of Keno Dam, Oregon.

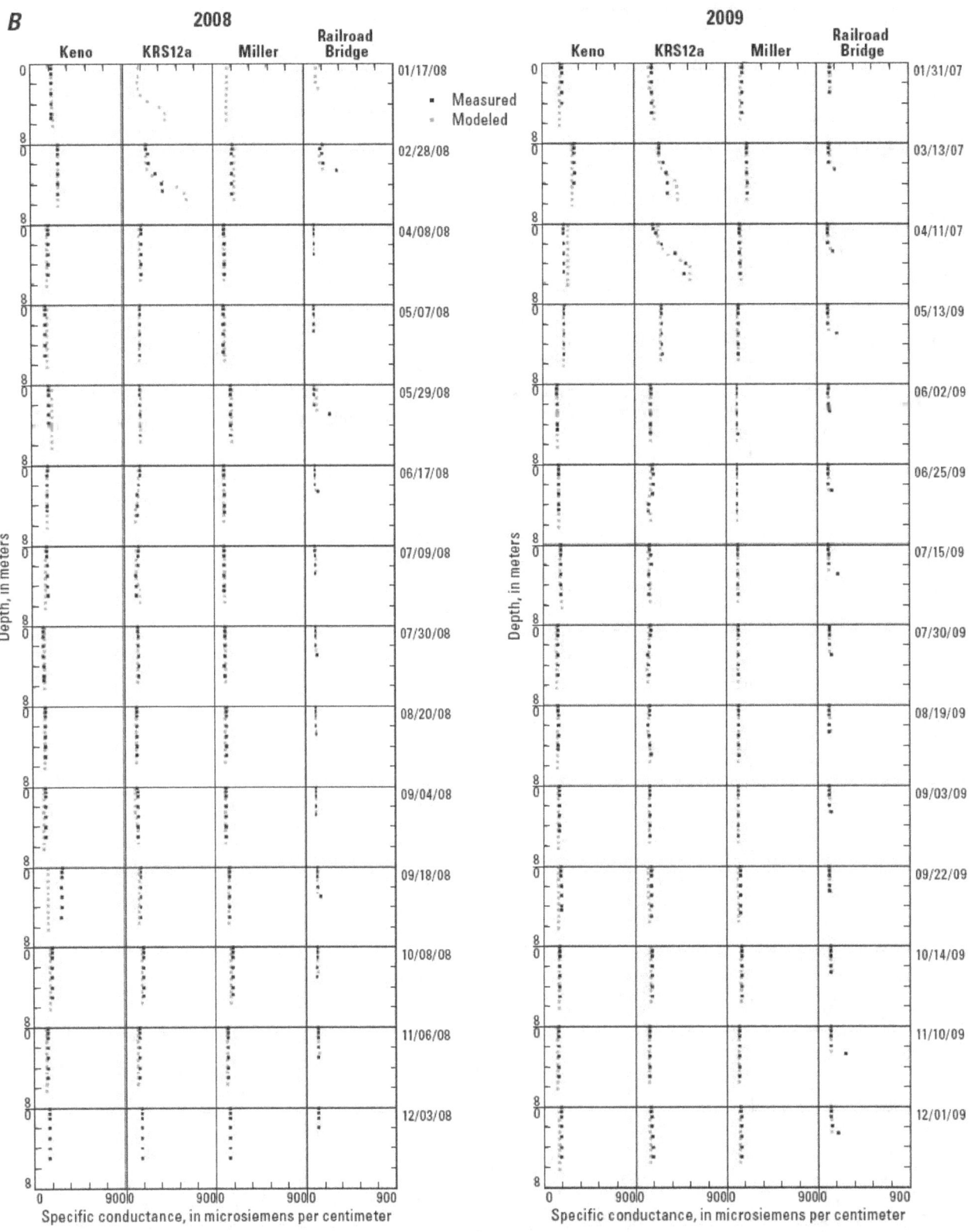

Figure 13.—Continued.

Phytoplankton

The presence, magnitude, and status of algal blooms are influenced by available light and nutrients, water temperature, travel time (flow), transport from upstream reaches, settling, mortality, and zooplankton grazing and other factors such as viruses. Algae can be beneficial because they are an important part of the food chain, and algal photosynthesis produces dissolved oxygen in the water column. Conversely, algal respiration and decay consumes dissolved oxygen. Some species of algae (for example, *Anabaena, Microcystis*) can produce toxins that may be dangerous to aquatic life or humans.

By simulating the algal community in the Link–Keno reach with three distinct types of algae (blue greens, diatoms, and other), the model was able to reproduce most of the major spatial and temporal trends in algae observed in 2006–09 (fig. 14). Diatoms typically bloomed in the spring, and large blooms of blue-green algae entered the reach from Upper Klamath Lake through Link River in summer (note that the maximum scale on the blue-green algae graphs in figure 14 is about 20 times higher than that for diatoms or other algae). The model simulates the characteristic decrease in blue-green algae concentration that occurs with increasing distance downstream of Link River (Sullivan and others 2008, 2009). Populations of diatoms and other algae do not decrease in the downstream direction, and at certain times, increase in the downstream direction. It is unknown why the blue-green algae populations from Upper Klamath Lake are not able to sustain themselves to the same levels in this reach. Several explanations have been proposed, including physical cell damage as a result of transport past Link Dam, differences in the characteristics and vertical thermal structure of the hydrodynamic system of the lake versus the river, or algal mortality due to low dissolved-oxygen concentrations. The latter mechanism was hypothesized and invoked by a previous model of this reach; that model separated algae into a "healthy" or, upon exposure to low dissolved oxygen, an "unhealthy" algae group. The two groups were assigned different growth, respiration, excretion, and mortality rates (Tetra Tech, 2009; Rounds and Sullivan, 2009, 2010). However, at present, insufficient evidence is

available to support a specific mechanism for the decline in algal populations through the Link–Keno reach. The model described herein does not simulate the details of this unknown process, but it does simulate the end result, which is an increase in settling and mortality losses for blue-green algae in this reach. If future research reveals the causal mechanism for blue-green algae losses in this reach, that process could be encoded into the model.

The CE-QUAL-W2 model keeps track of whether light, nitrogen, or phosphorus is limiting algal growth in each model cell for each time step. That information showed that light was the major limiting factor to algal growth, due to large light extinction coefficients that result from the dark color and prevalence of dissolved organic matter and suspended matter. The model did indicate that nutrients were the major limiting factor for algal growth for some periods, locations, and depths; this included occasional phosphorus limitation for the blue-green algae group and nitrogen and phosphorus limitations for the other two algal groups when sufficient light was available.

Models used to simulate algal communities, even when multiple algal groups are separated and simulated separately as in this study, can only begin to capture the complexity of those communities and their interactions and responses to light, flow, vertical mixing, nutrients, and dissolved oxygen. Unlike heat-exchange processes, the model algorithms used to represent algal communities are gross simplifications of actual processes, and large uncertainties in the predictions are inherent. Sources of error for algae in this model include uncertainties and variations in almost all of the growth rates, reaction rates, settling rates, and stoichiometric factors. In addition, errors are inherent to the estimation of Link River algal inflow concentrations in 2006 and 2009 as well as short-term variation in the Link River algal concentration that was not captured by weekly grab sampling in 2007 and 2008. Fewer algal data existed for tributaries such as the Lost River Diversion Channel and Klamath Straits Drain. Furthermore, CE-QUAL-W2 does not include an algorithm to account for blue-green algae changing their buoyancy as a function of their physiological state. Despite these uncertainties and limitations, the major spatial and temporal patterns for algae were well-simulated by the model, which suggests that most of the processes affecting algae were simulated with sufficient accuracy to be predictive and useful.

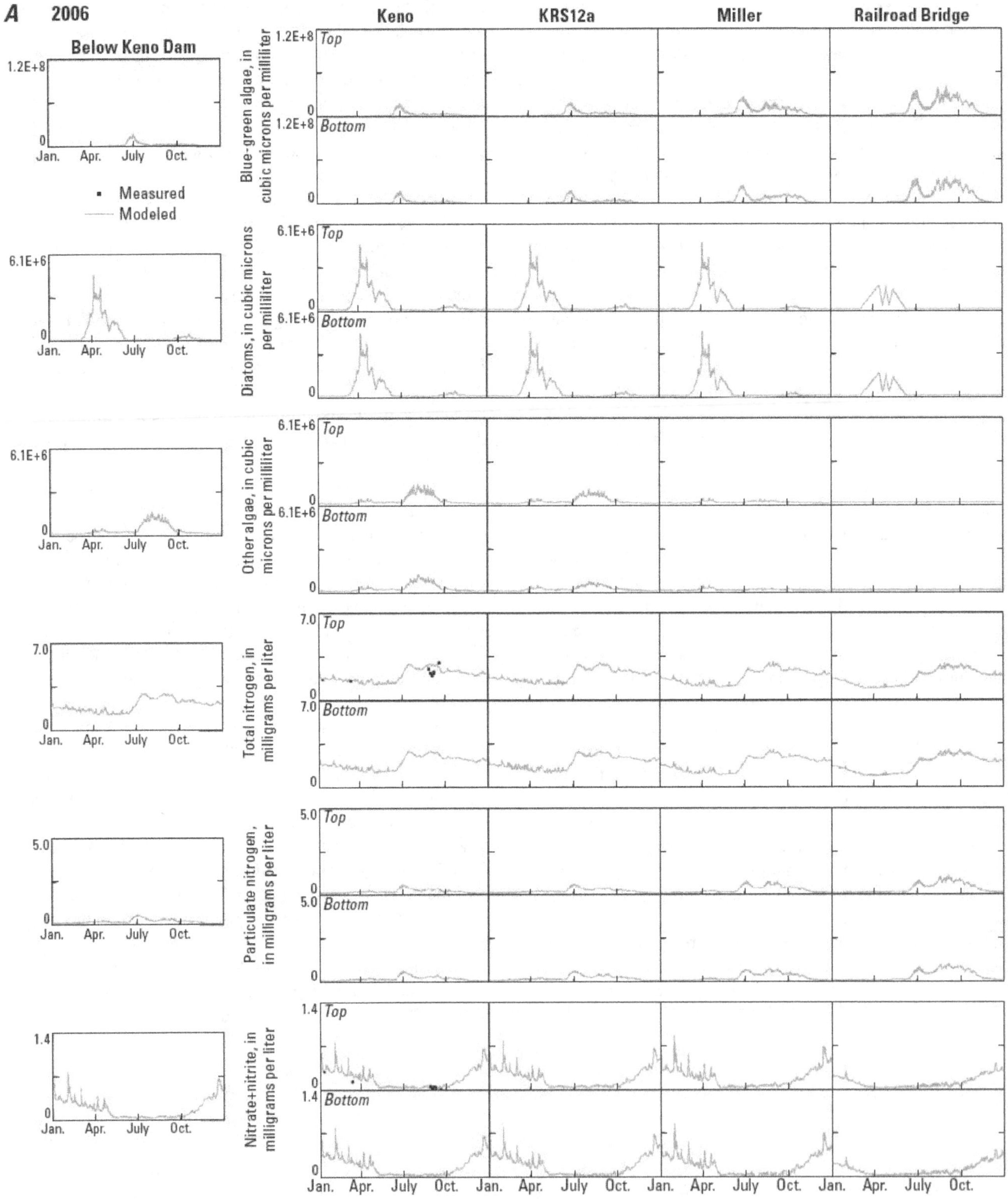

A   2006

Figure 14. Measured and modeled algae, nutrients, and organic matter for calendar years (*A*) 2006, (*B*) 2007, (*C*) 2008, and (*D*) 2009 for sites in the Klamath River upstream of Keno Dam, Oregon.

A   2006—Continued

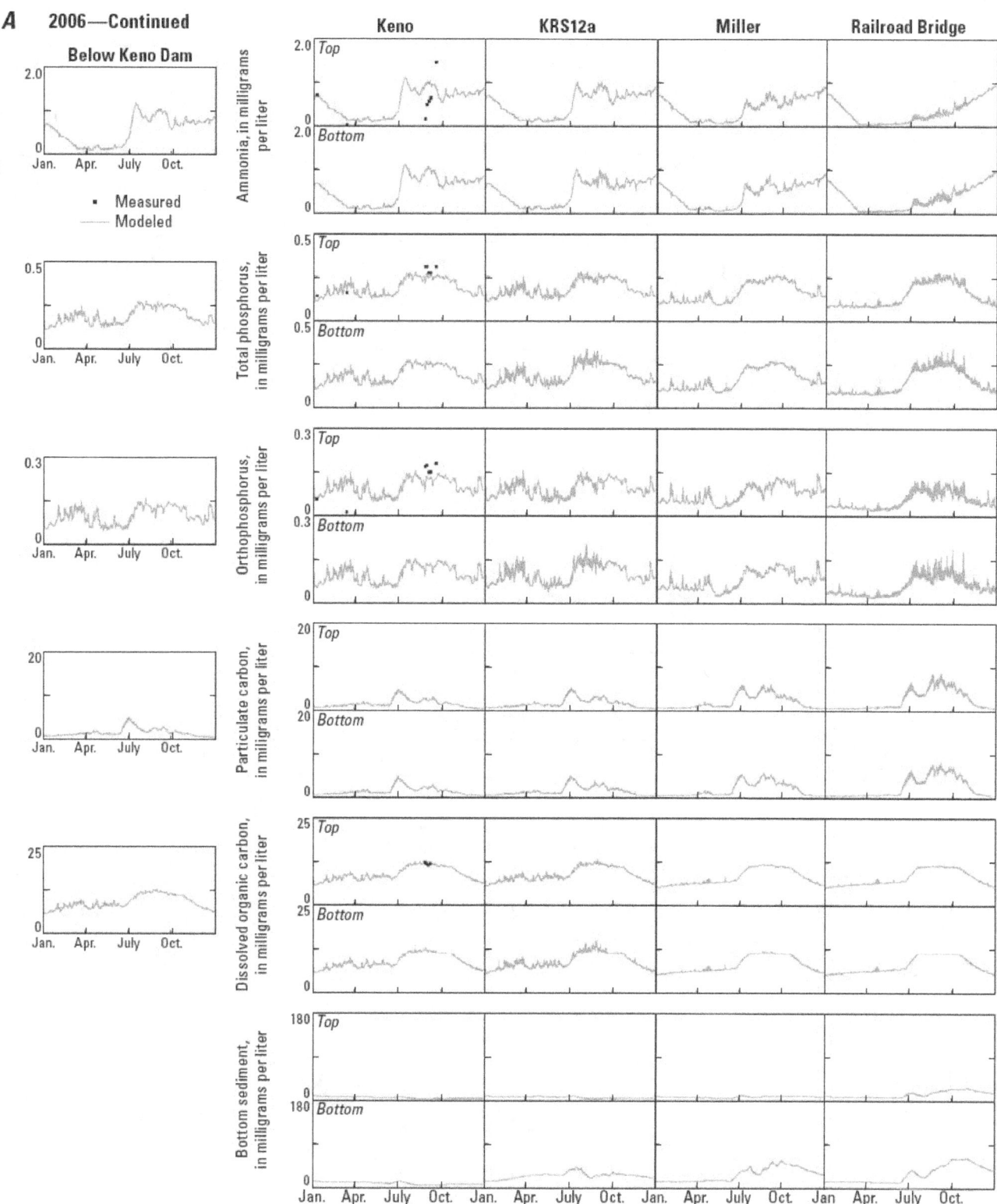

Figure 14.—Continued.

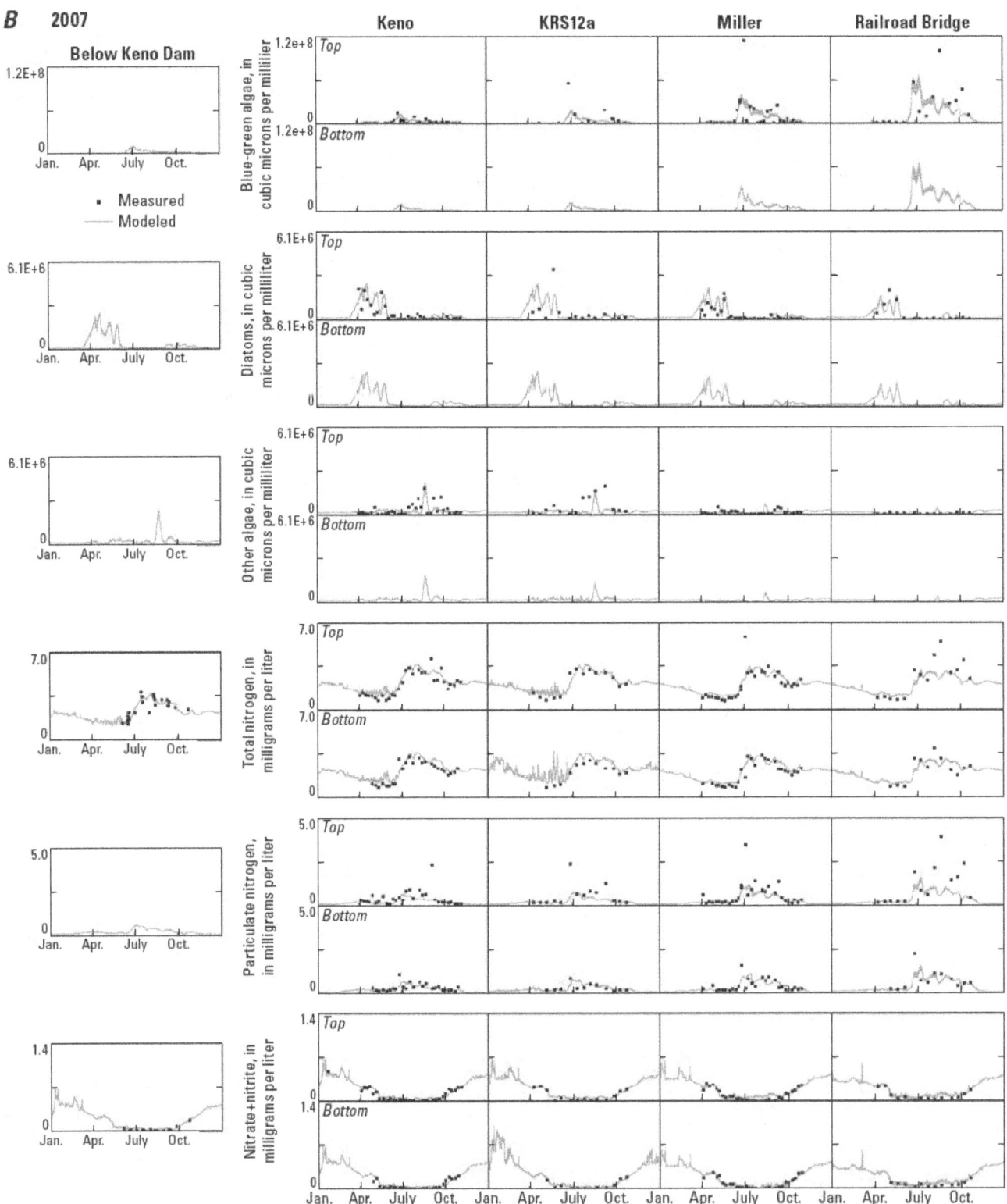

Figure 14.—Continued.

B   2007—Continued

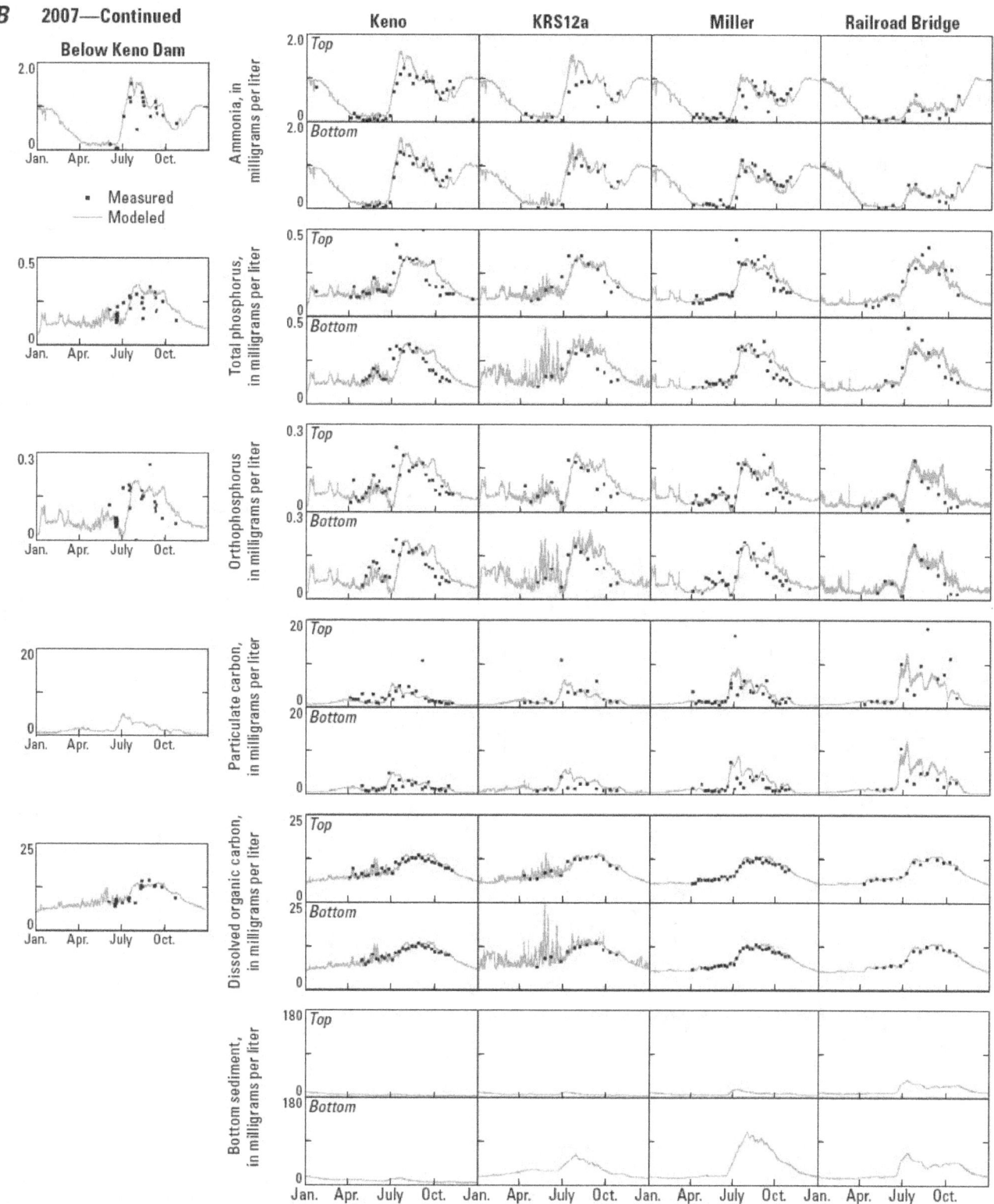

Figure 14.—Continued.

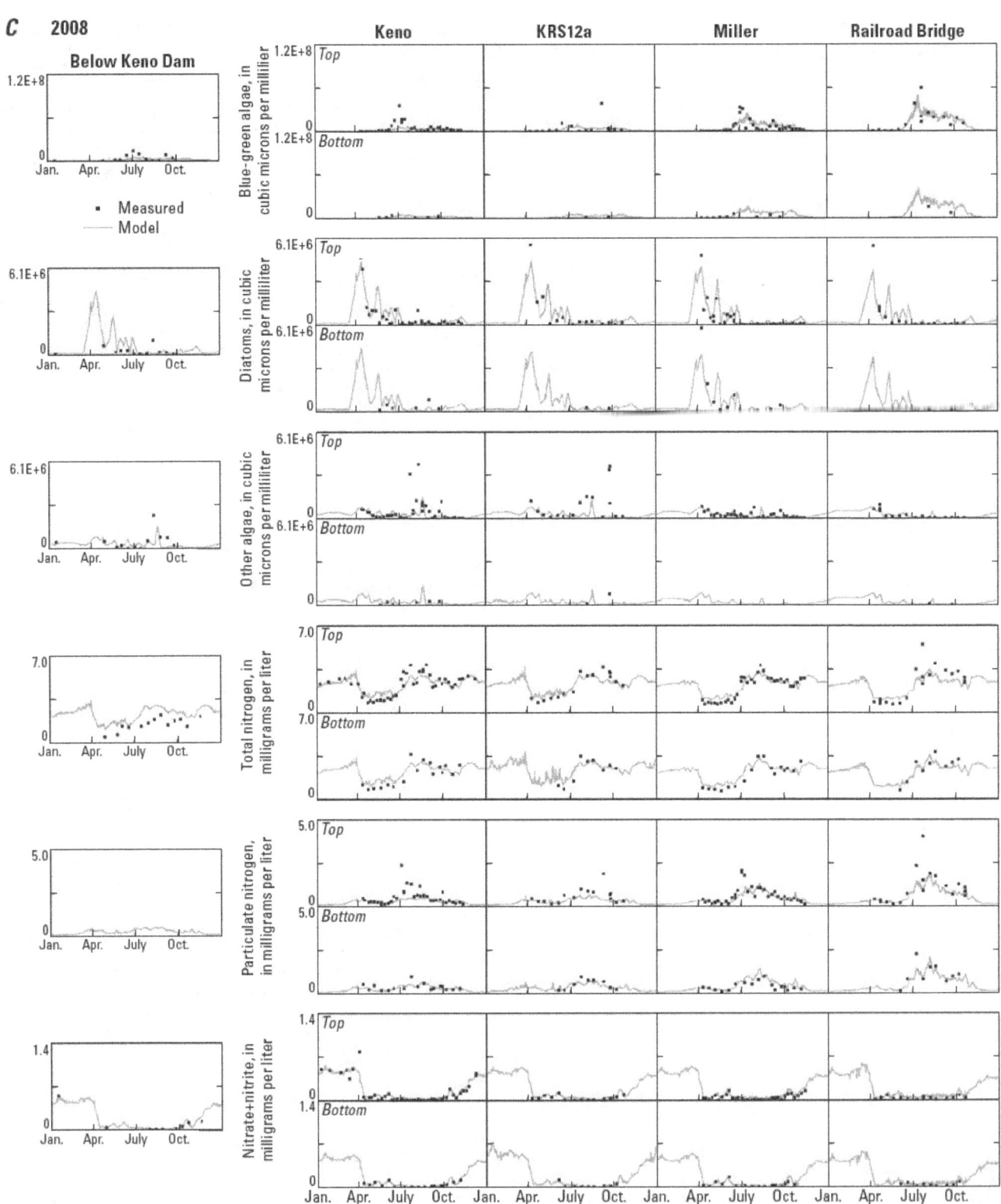

Figure 14.—Continued.

C   **2008—Continued**

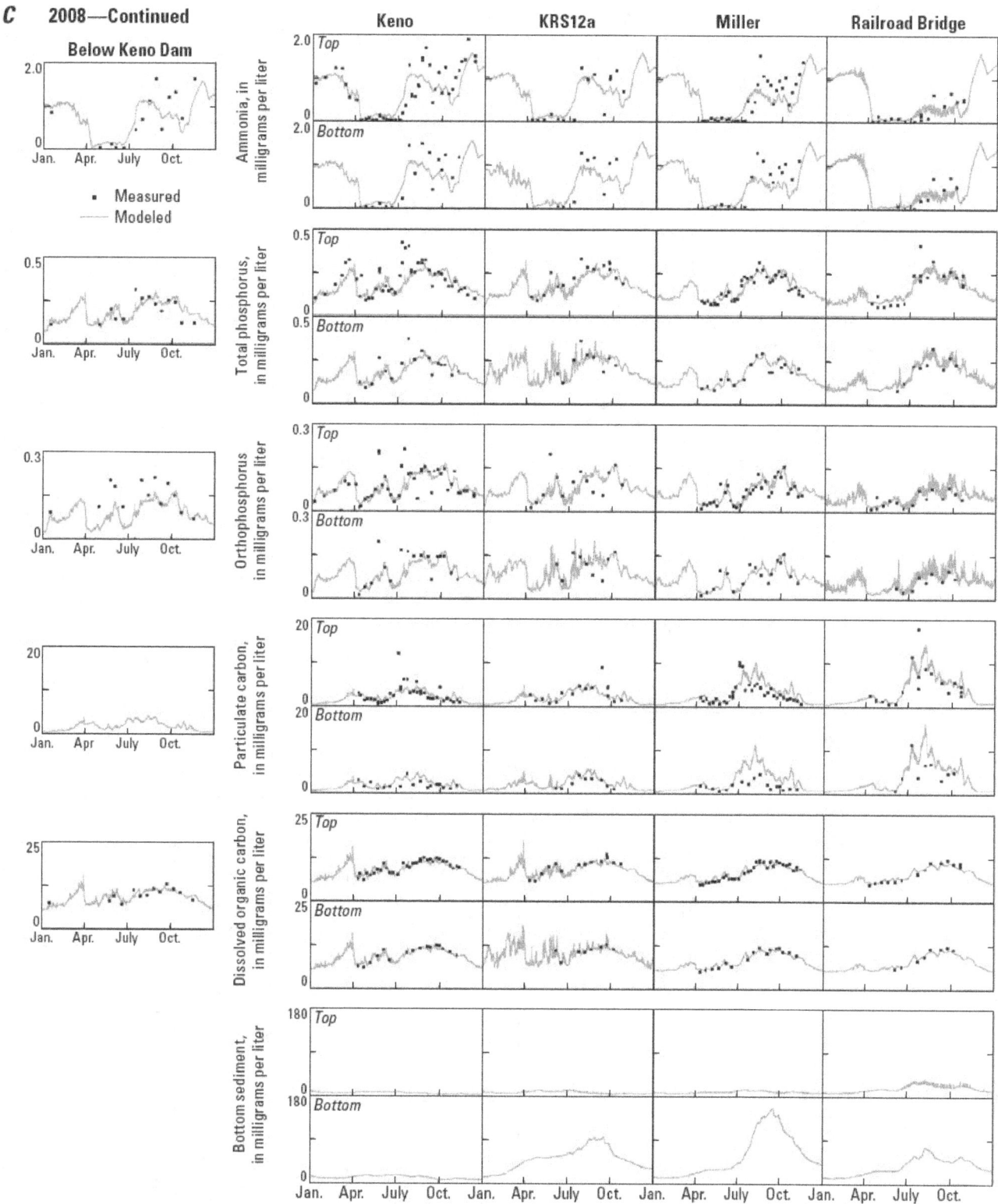

Figure 14.—Continued.

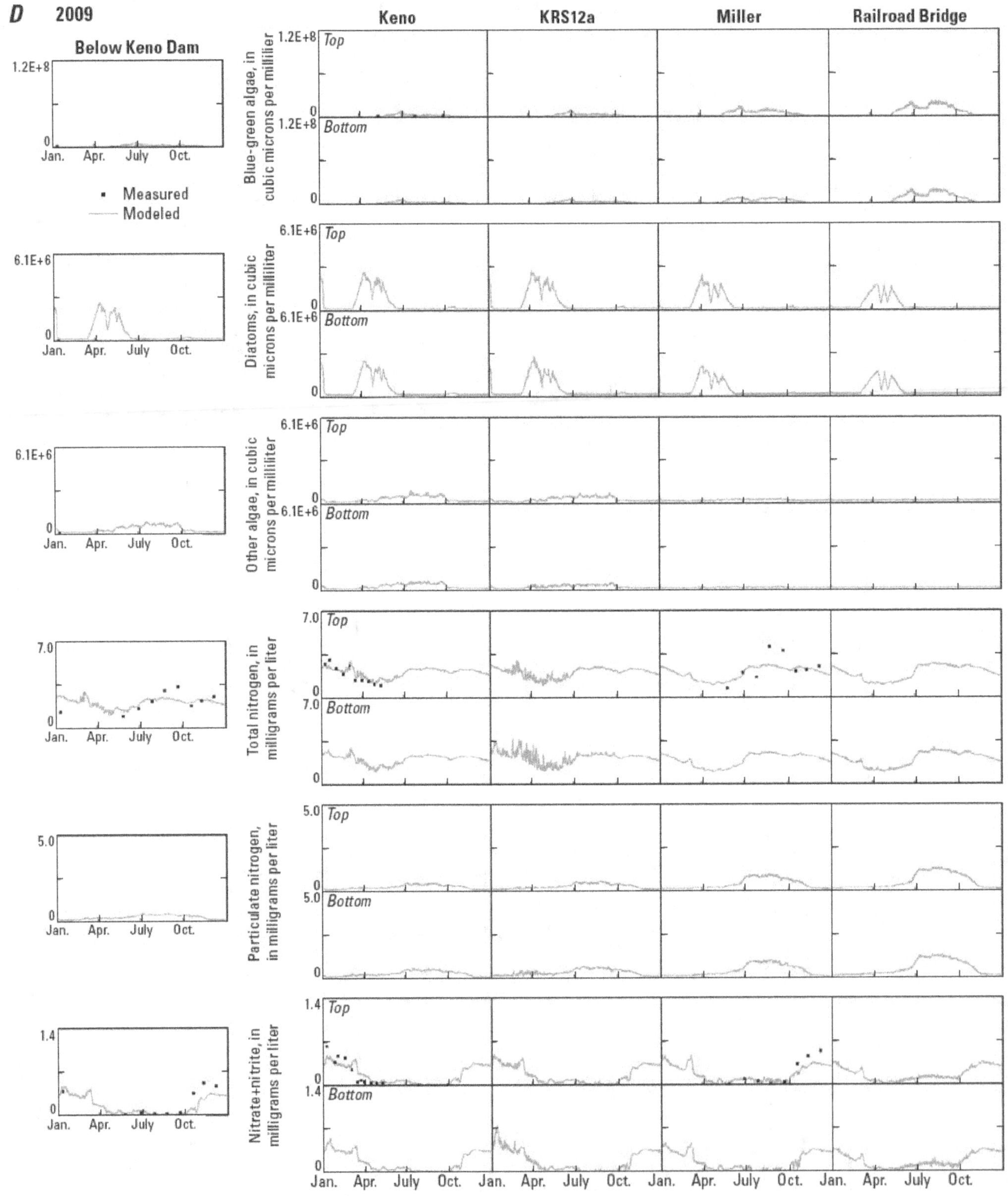

Figure 14.—Continued.

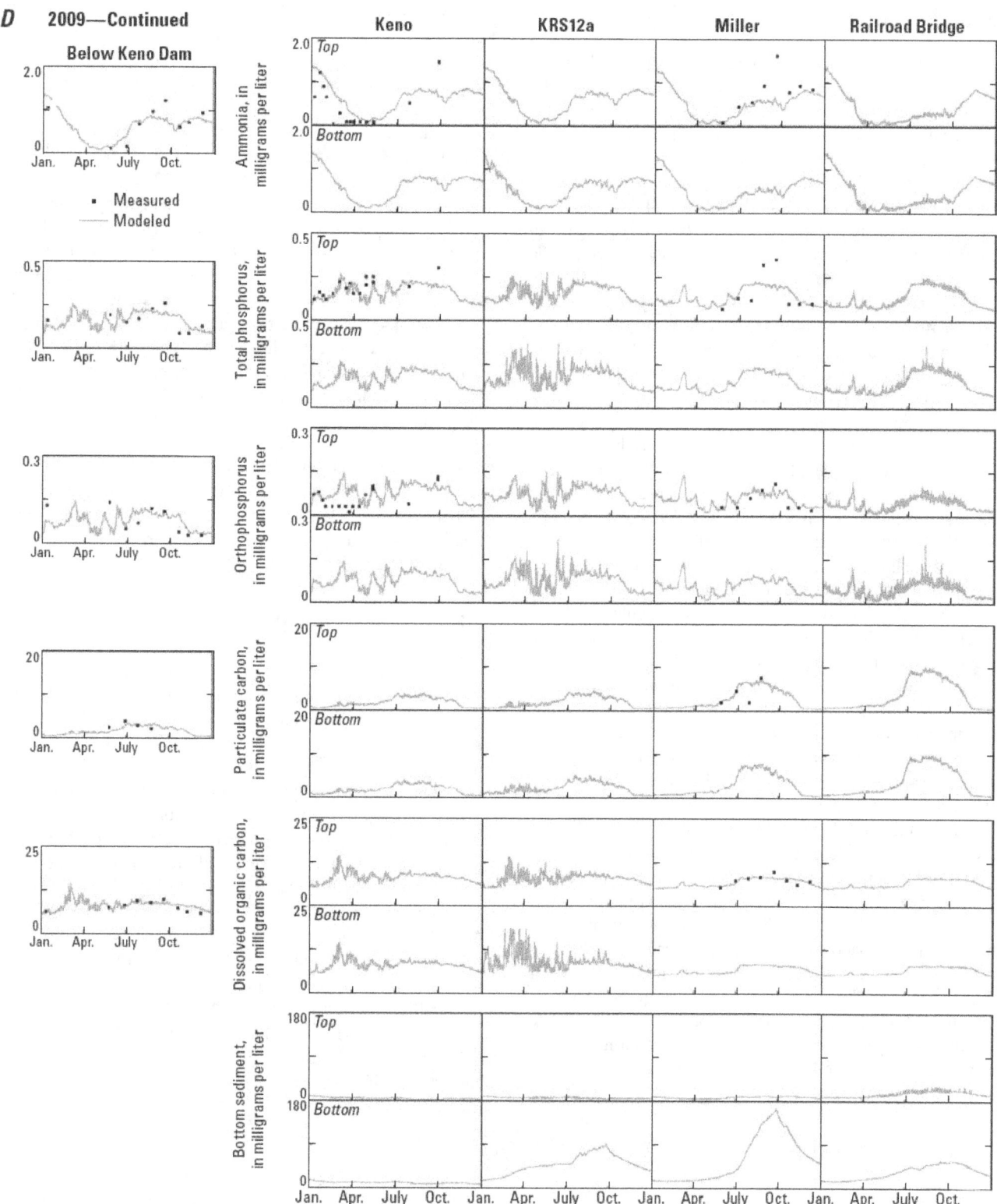

D   **2009—Continued**

Figure 14.—Continued.

Organic Matter

Organic matter is of paramount importance to the water quality of the Link–Keno reach, and its cycling is closely tied to the concentrations and dynamics of regulated constituents such as nutrients and dissolved oxygen. Particulate organic matter is composed of a diverse collection of material such as dead algae, zooplankton, leaf litter, bacteria, and other organic materials in various stages of decay. Dissolved organic material is made up of organic molecules of varied origin and composition, typically with complex chemical structures that defy exact characterization.

The composition, and therefore the decay rates, of organic matter varies widely. Quickly decaying (labile) organic matter can result in the consumption of large amounts of oxygen over short time periods on the order of hours to days, with a concomitant release of dissolved nutrients. In contrast, slowly decaying (refractory) organic matter causes less oxygen to be consumed over short periods, but can continue to cause oxygen loss and release of nutrients over longer periods (and over longer downstream distances) when more labile materials might have been exhausted.

Long-term (30-day) BOD tests of water in the Link–Keno reach have shown the presence of both labile and refractory forms of organic matter; dead algal material was especially labile, and dissolved organic matter was particularly refractory (Sullivan and others, 2010). These results for the Link–Keno reach are similar to observations reported for other systems in the scientific literature. For example, in another eutrophic system, algal material was determined to be the most labile substrate and dissolved organic matter the most refractory; treated wastewater treatment plant effluent and pulp and paper waste were of intermediate lability in that system (Hendrickson and others, 2007).

Although many data were available to characterize the lability of organic matter in this reach in summer, less is known about the character of organic matter in winter. Degradation of organic matter is fastest soon after death of the source material (Canuel and Martens, 1996), so winter organic matter is likely made up of older, more recalcitrant organic matter. Organic matter was assumed to be less labile in winter than in summer when abundant algae were present. Model sensitivity testing confirmed that to correctly simulate winter dissolved-oxygen concentrations, organic matter must be more refractory and less labile in winter than in summer.

Organic matter can decay either in the water column or on the river bottom. The amount and decay characteristics (labile or refractory) of organic matter that falls to the river bottom is tracked by the model (fig. 14, plots of "Bottom sediment") and is used to produce an appropriate and seasonally responsive level of SOD. The organic material in the first-order sediments compartment in the models built up and was mostly consumed within a year. This representation has implications for management of water quality in this reach, and suggests that if the materials that most contribute to first-order SOD (algae and LPOM, especially from Upper Klamath Lake) are removed from the system or decreased in load, oxygen levels could increase significantly.

The model simulated the seasonal and spatial patterns of particulate carbon and dissolved organic carbon well, as evidenced by comparisons to weekly grab samples (fig. 14, table 6). Measured particulate carbon includes algae and POM, as does model output for particulate carbon. Given the typical concentrations of POC and DOC (0.5–18 mg/L and 5–14 mg/L, respectively) in the Link–Keno reach, these error statistics (MAE of about 1.4 and 0.7 mg/L for POC and DOC, respectively) indicate that model errors typically were small relative to the amount of dissolved and particulate carbon moving through the system.

Nitrogen and Phosphorus

Nitrogen and phosphorus are essential nutrients contributing to river primary productivity. Dissolved nitrate and ammonia are the most commonly measured nitrogen species, along with total nitrogen, which includes nitrogen in organic matter, both dissolved and particulate. Nitrite can occur in natural waters, but measurements showed its concentration to be low in this reach (Sullivan and others, 2008, 2009); therefore, nitrite results are combined with nitrate results in this discussion. Sources of nitrate in the model include ammonia nitrification, and sinks include denitrification and algal uptake. Model sources of ammonia include sediment release, algal respiration, and organic matter decay; sinks include nitrification and algal uptake. Tributary inflows and outflows also affect concentrations.

Nitrate concentrations were highest in winter and lowest in summer; this temporal pattern was common and was observed at the Link River inflow and elsewhere in the Link–Keno reach at most sites and years (fig. 14). The model simulated the temporal and spatial patterns of nitrate well. Like nitrate, ammonia concentrations were relatively high in winter. Unlike nitrate, ammonia concentrations increased in midsummer from relatively low concentrations in early summer. Ammonia concentrations showed a distinct spatial pattern in summer. Summer concentrations were less than 0.1 mg/L in the Link River inflow, but usually increased to greater than 1 mg/L downstream at Keno. Model fluxes showed that the source of most ammonia was from the decay of organic matter in the sediments and water column with algal processes as an additional source. Concentrations of un-ionized ammonia (NH_3 as opposed to the ionized form NH_4^+), which has high toxicity to aquatic life, also increased through the reach in summer (fig. 15).

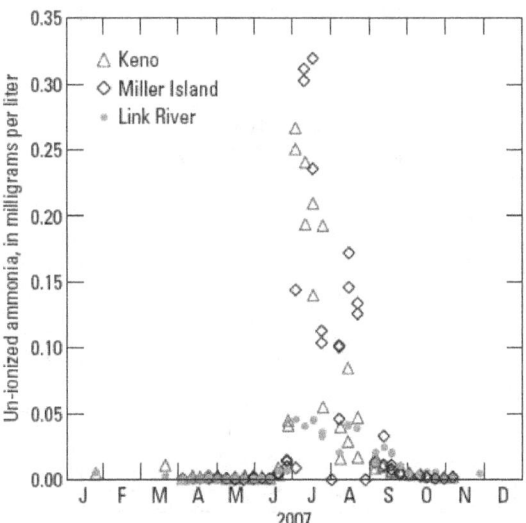

Figure 15.   Un-ionized ammonia concentration calculated from measured data for 2007 at the Link River, Miller Island, and Keno sites, Oregon.

Phosphorus exists in the dissolved phase as orthophosphorus, and also as a part of dissolved organic matter. Measurements of total phosphorus include orthophosphorus adsorbed to inorganic particles (not simulated in this model) and amounts contained in particulate organic matter and algae. Sources of orthophosphorus in CE-QUAL-W2 include various inflows, algal respiration, anoxic sediment release, and organic matter decay; sinks of orthophosphorus include algal uptake. Spatial and temporal patterns of total phosphorus and orthophosphorus were simulated well by the model (fig. 14). Orthophosphorus patterns were matched well by the model in spring, but the model sometimes overpredicted the concentration slightly in the fall. This could indicate that there may be some seasonal differences in the stoichiometric ratios for phosphorus, whereas the model allows only one ratio to be set for the whole year. Given the typical concentrations of these constituents in the river, the overall nitrogen and phosphorus error statistics (table 6) indicate that model errors were relatively small compared to the amount moving through the system.

Modeled loads of total nitrogen and total phosphorus at the reach inflow at Link River and at the reach outflow at Keno Dam (fig. 16) showed differences among years, with the highest loads in the high flow year 2006. This difference is reasonable, since loads are calculated by multiplying flow by concentration. Some consistent seasonal patterns were present. Higher nutrient loads were exported downstream compared to incoming loads from Upper Klamath Lake for the period February through April, a result that likely is due to loads entering from tributaries. However, nitrogen and phosphorus loads exported downstream were lower than the respective loads imported from Upper Klamath Lake for the summer period of July and August. These lower loads are due, in part, to less flow entering the river in summer and greater withdrawals, compared to winter in addition to particulate-associated settling losses in summer.

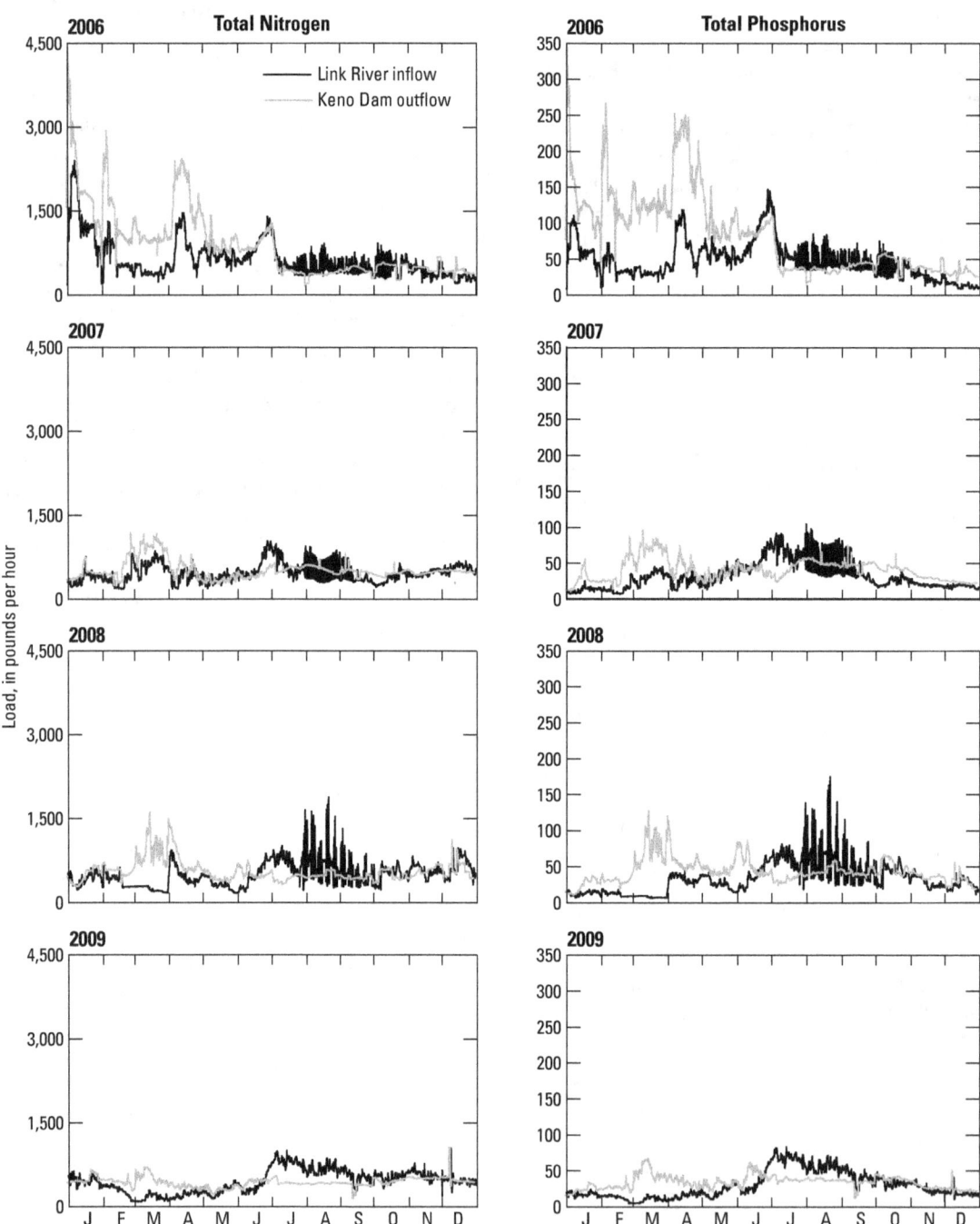

Figure 16.   Simulated hourly loads of total phosphorus and total nitrogen at the Link River inflow and the Keno Dam outflow, Oregon, calendar years 2006–09.

Dissolved Oxygen

Fish and most other aquatic organisms need suitable levels of dissolved oxygen to survive and function. In the model, sources of dissolved oxygen included the inflows, exchange with the atmosphere across the river surface, and phytoplankton photosynthesis. Sinks of dissolved oxygen included atmospheric exchange (when concentrations were supersaturated), algal respiration, ammonia and nitrite oxidation, and organic matter decay both in the water column and sediments. The dissolved-oxygen budget is complex and dynamic, and can be particularly difficult for a model to accurately simulate when algal communities and oxygen demands are large.

A plot of all 4 years of measured dissolved-oxygen data against water temperature data from the Miller Island (top) site reveals a repeating annual cycle that reflects the effects of temperature (oxygen solubility), photosynthesis, and large oxygen demands becoming predominant at different times of the year (fig. 17). Figure 17 includes contour lines of constant percent oxygen saturation (as a function of temperature), with those lines indicating saturation and supersaturation (solid contour lines) or subsaturation (dashed contour lines). Data collected during different months are plotted in different colors to illustrate the annual cycle. In January and February, dissolved-oxygen concentrations are near saturation with only minor daily variations. In March through June, the system is warming and algae are actively producing oxygen through photosynthesis because supersaturated conditions in excess of 150 percent occur. In July or August, the algal community cannot produce enough oxygen to counteract the oxygen demands from respiration, BOD, and SOD, and the oxygen levels decrease to values near zero at times. For the rest of the summer and into early fall, photosynthesis continues to decrease and oxygen demands remain high; oxygen levels slowly recover toward saturation as the labile oxygen demands are exhausted and the water cools in late fall and into winter.

The model was able to simulate most of these patterns of dissolved oxygen in the river, both near the surface and at depth (fig. 18). A notable difference was that the measured data showed large daily variations in near-surface dissolved-oxygen concentrations along most of the river, but the model simulated large daily variations only in the upstream river reaches, not at the KRS12a or Keno near-surface sites. The model responds to field observations and data indicating that algal biovolume usually decreased longitudinally from Link River to Keno. It is not known with certainty why the near-surface measured data continued to show large daily oxygen cycles, even in the presence of fewer algae downstream. Observational evidence indicated that rooted aquatic plants (macrophytes) may be more common in the downstream portions of this reach; they may exist in large enough quantities that their photosynthesis and respiration patterns were contributing substantially to the near-surface

daily oxygen cycles there. Macrophytes are not yet included in this Link–Keno model, although CE-QUAL-W2 does include algorithms to simulate macrophyte populations (Cole and Wells, 2008).

An analysis of the modeled oxygen fluxes for a representative year (fig. 19) indicated that besides algal photosynthesis, reaeration from the atmosphere was an important source of oxygen in fall, when hypoxic or anoxic conditions in the river caused a large oxygen deficit relative to its solubility and, therefore, a large gradient for absorbing atmospheric oxygen. The reaeration flux also showed that there were short periods, during supersaturation due to algal photosynthesis, when the river outgassed oxygen to the atmosphere. Sinks of oxygen, as expected, were dominated by decay of algae and organic matter primarily on the channel bottom and in the water column. Algal respiration and ammonia nitrification were smaller sinks of dissolved oxygen in this reach.

Simulated total SOD, the sum of first-order and zero-order demand, varied both spatially and temporally. For example, from January through mid-June and late December 2008, modeled SOD was less than 1.5 (g O_2/m^2)/d (fig. 20). Modeled SOD was elevated in summer and fall 2008 with values up to about 15 (g O_2/m^2)/d. Elevated SOD was due to warmer temperatures and to the settling and decay of labile particulate organic matter and blue-green algae. An *in-situ* SOD study for the Link–Keno reach took place in early June 2003, before the summer influx of algae and organic matter (Doyle and Lynch, 2005). Those early June SOD measurements ranged from 0.3 to 2.9 (g O_2/m^2)/d, with a median of 1.8 (g O_2/m^2)/d, values in line with early June SOD from the model. *In-situ* SOD measurements require oxygen to be present in the water column and so cannot be made during the anoxia that was present through much of the reach in summer. However, summer *in-situ* SOD greater than 10.2 (g O_2/m^2)/d has been measured in Upper Klamath Lake (Wood, 2001). A 1-in. thick layer of "fluffy" material topped the sediments during that measurement and it was theorized that the source of this material was settled algal mats.

The actual exerted SOD was less than the full potential SOD for times or locations where the river was anoxic. For example, the highest simulated SOD in July and August 2008 was near Link River (fig. 20), because enough dissolved oxygen was present in the water column to support the full expression of SOD. Substantial deposition of organic particulate materials occurred both there and downstream, but SOD could be expressed only when oxygen was present; anoxia prevents the expression of additional SOD. The organic sediment continued to build up, however, and this led to periodic spikes in the simulated SOD rate in September, when dissolved-oxygen concentrations increased and allowed SOD to be expressed, as long as decaying organic matter was still present in the sediment.

Klamath River at Miller Island Boat Ramp [top] (420853121505500)

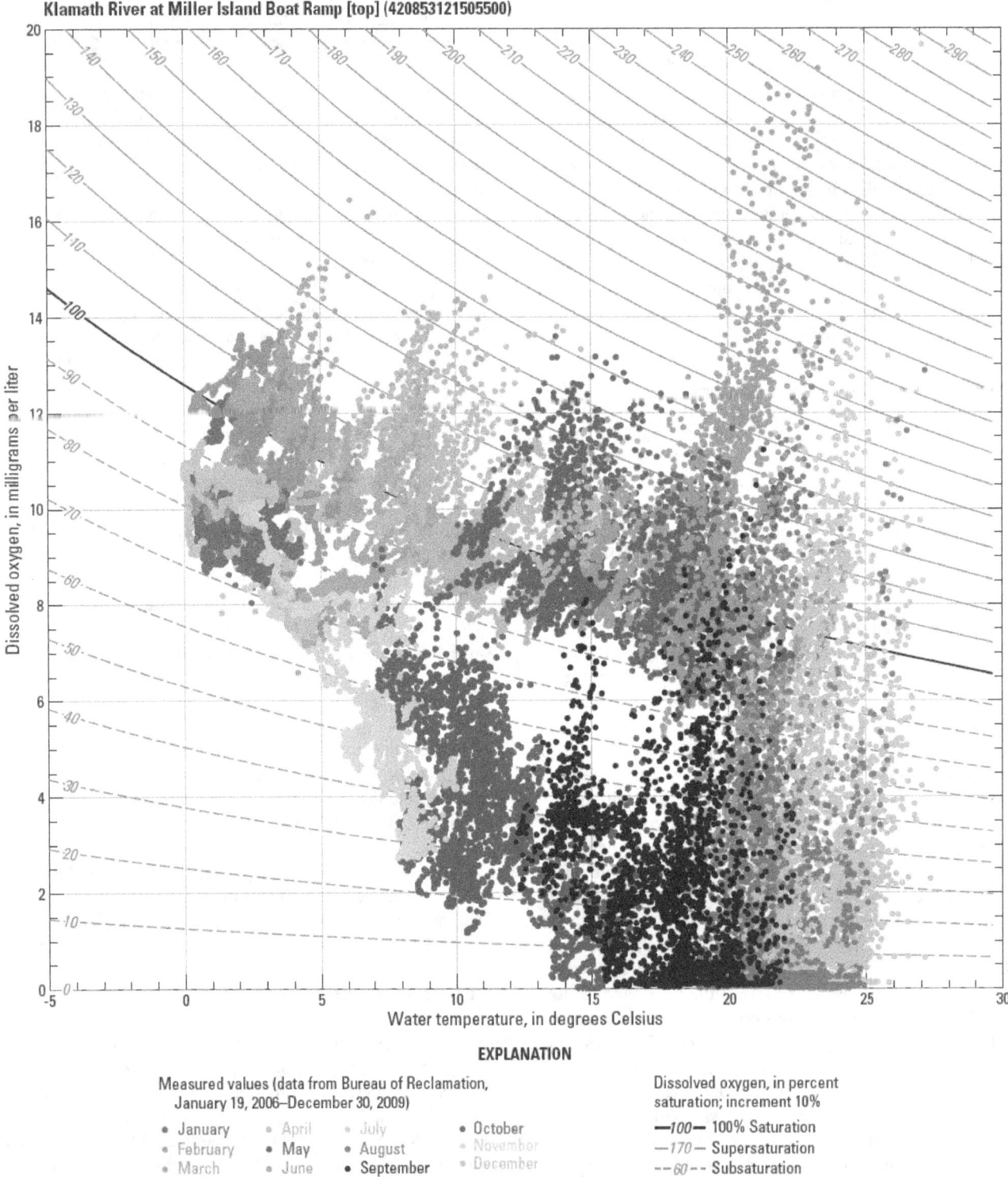

EXPLANATION

Measured values (data from Bureau of Reclamation,
January 19, 2006–December 30, 2009)

• January	• April	• July	• October
• February	• May	• August	• November
• March	• June	• September	• December

Dissolved oxygen, in percent
saturation; increment 10%

—*100*— 100% Saturation
—*170*— Supersaturation
--*60*-- Subsaturation

Figure 17.   Annual cycle of dissolved-oxygen concentrations at the Klamath River at Miller Island Boat Ramp site, Oregon (U.S. Geological Survey station 420853121505500, top). Concentrations are near saturation early in the year, then supersaturated from March through July, owing probably to photosynthesis. Dissolved-oxygen concentrations decrease to low levels in mid- to late summer as a result of large oxygen demands and slowly recover as summer ends and fall leads to cooler temperatures.

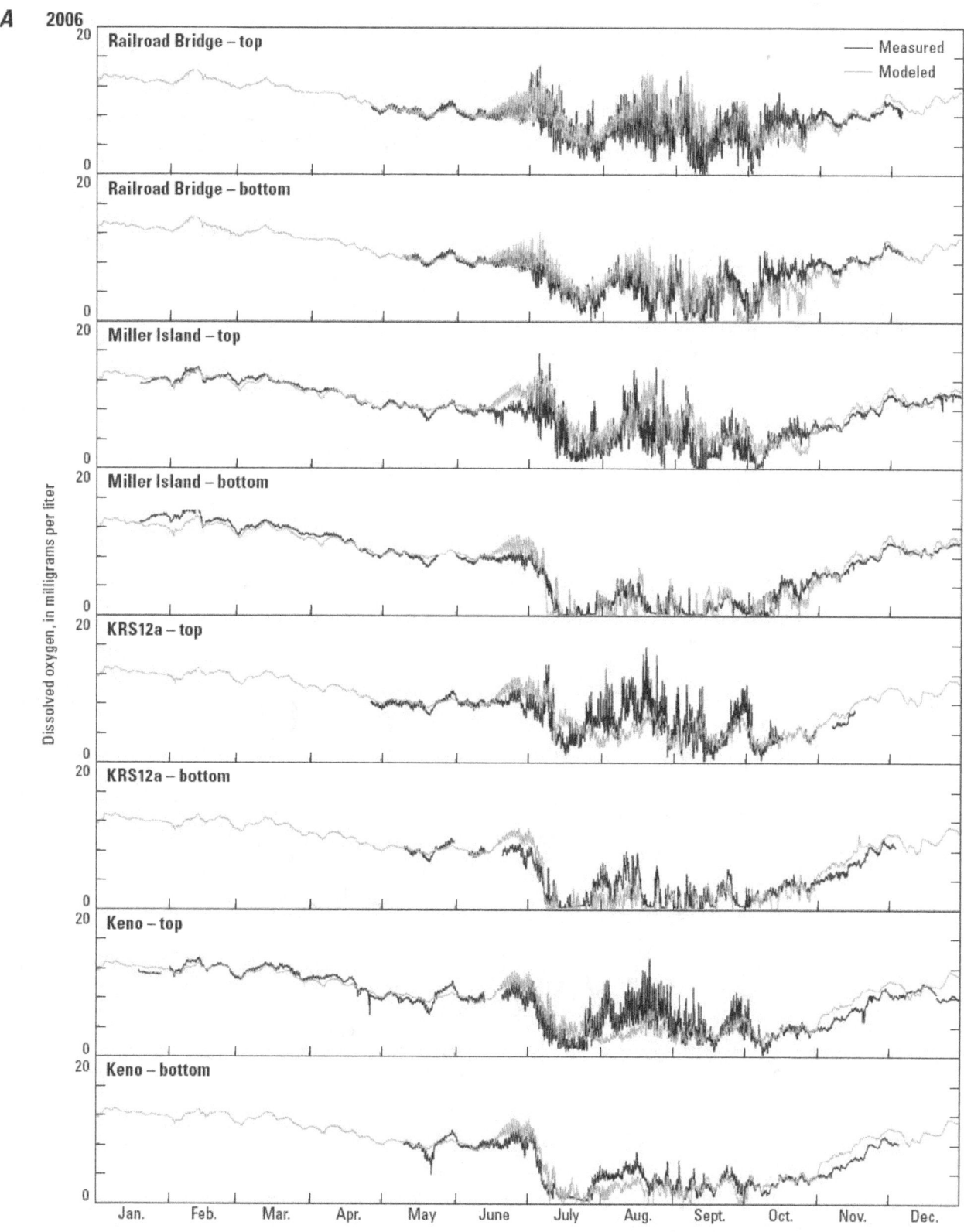

Figure 18. Measured and modeled hourly dissolved-oxygen concentrations during calendar years (*A*) 2006, (*B*) 2007, (*C*) 2008, and (*D*) 2009 for sites in the Klamath River upstream of Keno Dam, Oregon.

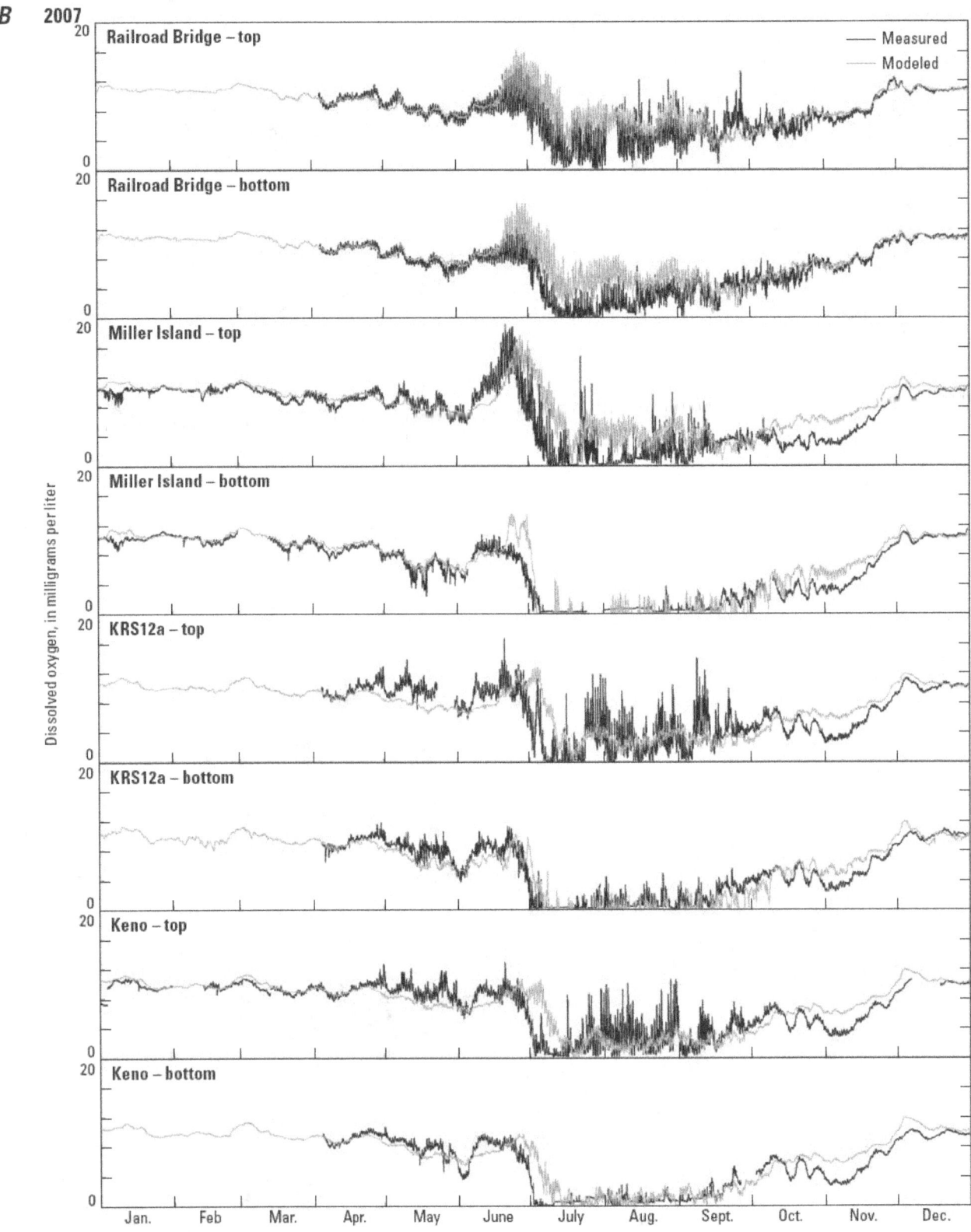

Figure 18.—Continued.

C   2008

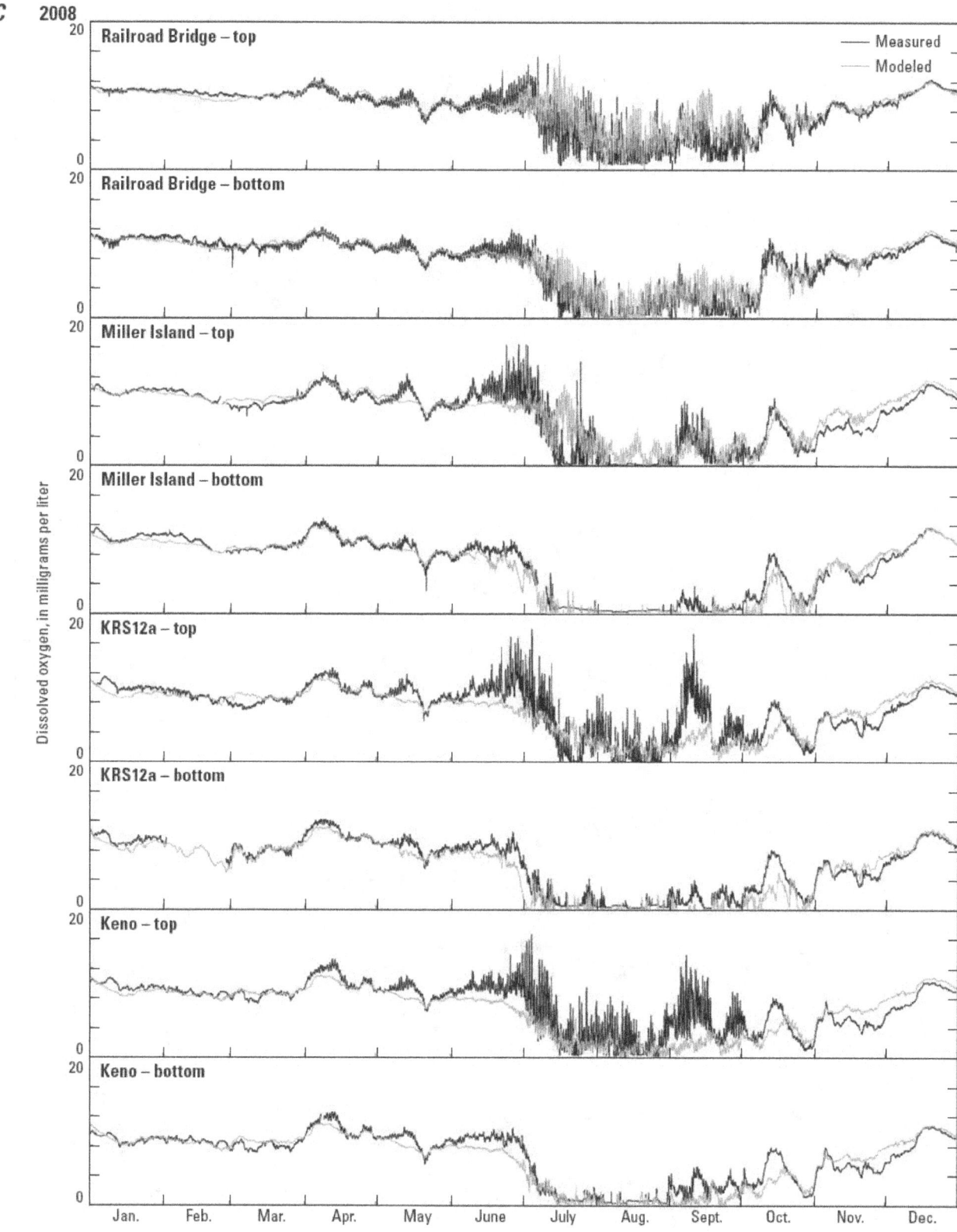

Figure 18.—Continued.

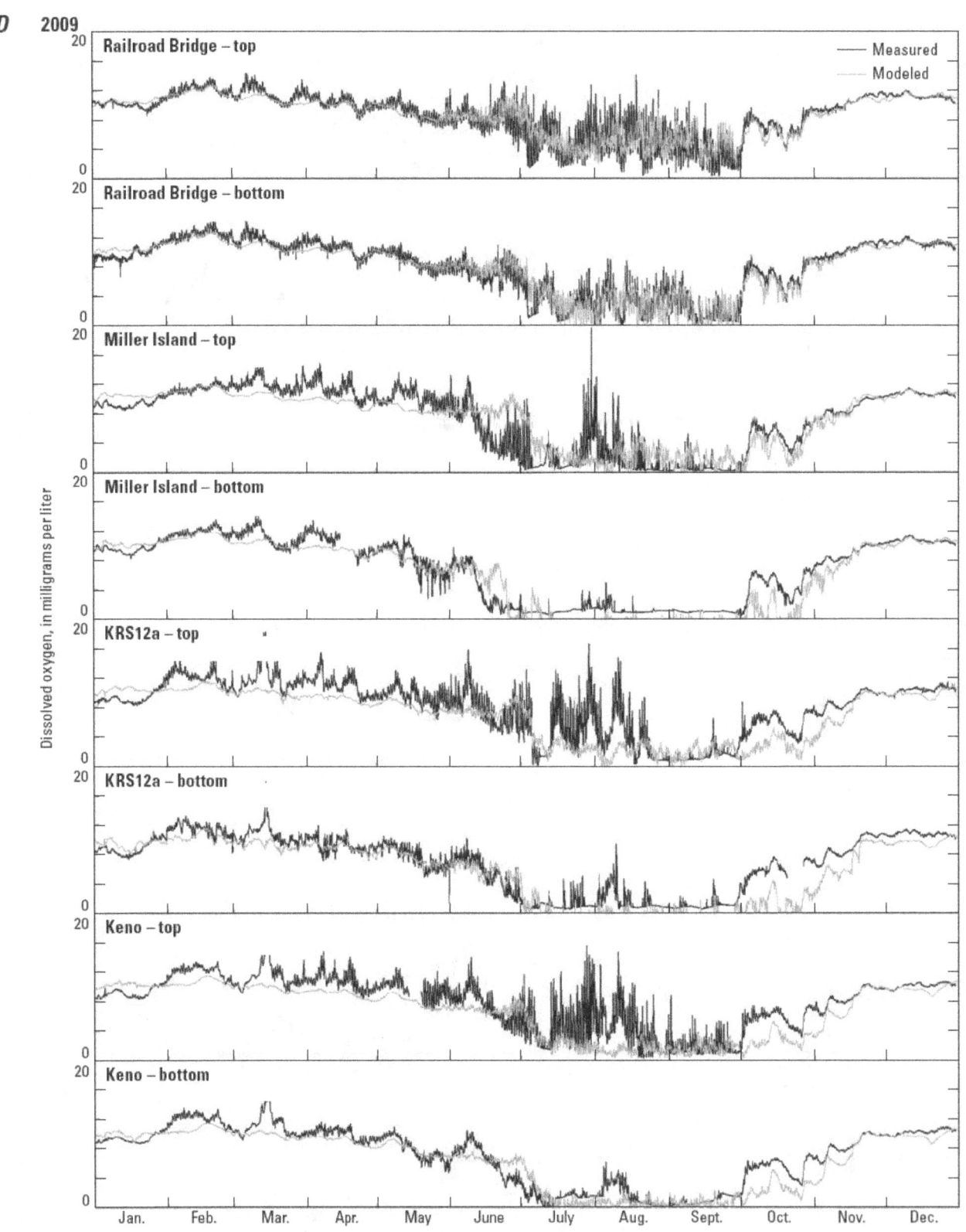

Figure 18.—Continued.

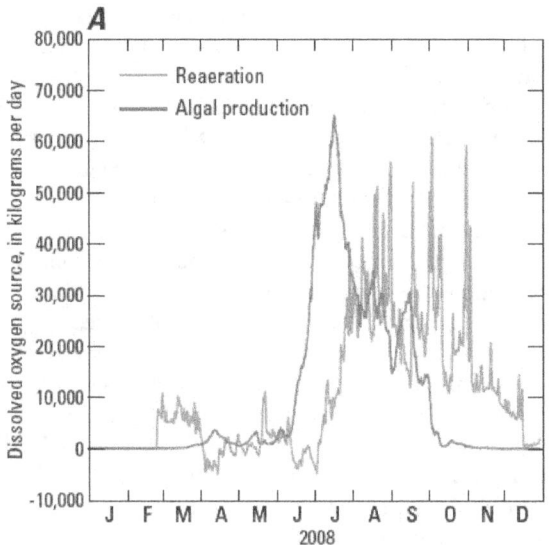
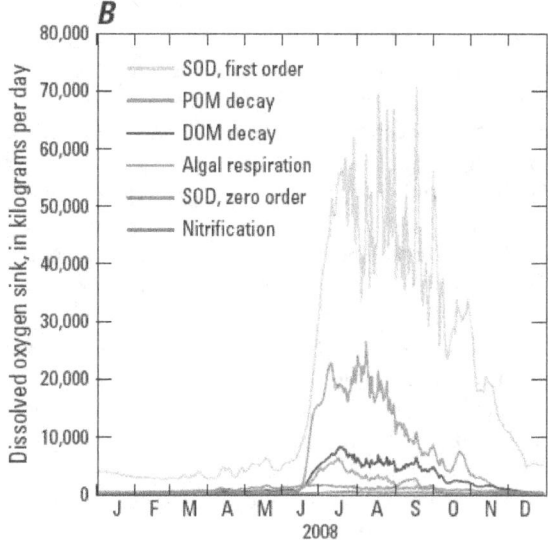

Figure 19. Simulated total annual dissolved-oxygen production (*A*, sources) and consumption (*B*, sinks) in the Link–Keno reach of the Klamath River, Oregon, calendar year 2008. DOM, dissolved organic matter; POM, particulate organic matter; SOD, sediment oxygen demand.

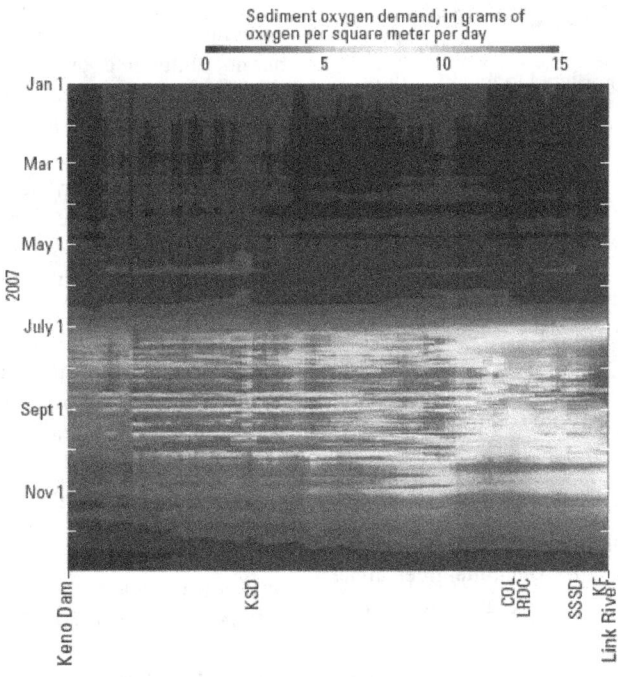

EXPLANATION

In-reach inflows

COL   Columbia Forest Products
KF   Klamath Falls wastewater treatment plant
KSD   Klamath Straits Drain
LRDC   Lost River Diversion Channel
SSSD   South Suburban wastewater treatment plant

Figure 20. Temporal and spatial distribution of modeled daily average sediment oxygen demand in the Klamath River between Link River and Keno Dam, Oregon, 2008. This plot represents total sediment oxygen demand.

The model does not currently simulate mechanical reaeration in Keno Dam releases, which results in an underprediction of dissolved-oxygen concentrations downstream of the dam during summer. If prediction of dissolved-oxygen concentrations downstream of Keno Dam becomes important, equations with empirical coefficients will need to be determined to relate dissolved-oxygen reaeration and spill rates (Cole and Wells, 2008); these equations will need to be developed specifically for Keno Dam. Reaeration equations also could be developed for use outside of the CE-QUAL-W2 model as a post-processing routine.

Goodness-of-fit statistics were calculated to compare hourly measured dissolved-oxygen data and model output at the same location and time (table 6). For 2006–09, the ME averaged over all sites for a year was between -0.66 and 0.51 mg/L. The MAE was between 1.01 and 1.30 mg/L. These results show relatively little bias and demonstrate that overall seasonal patterns were captured by the model. Some improvements might be possible through the simulation of macrophytes and a deeper understanding of some of the more-dominant factors and processes affecting dissolved oxygen.

Model Sensitivity and Uncertainty

A sensitivity analysis was completed to examine the response of changing selected model parameters to model results. The analysis was done by varying the value of the parameter of interest while holding all other model parameters constant. The Upper Klamath River model has a large number of parameters—enough that a formal sensitivity analysis was not performed on every parameter. The process of adjusting model parameters during calibration, in addition to the feedback from PEST runs, provided a general sense of parameter sensitivity. In addition, several formal sensitivity tests were performed to examine the effect of varying parameters by 20 percent (table 7). These parameters included the wind sheltering coefficient (WSC), light extinction coefficient for water and dissolved constituents (EXH2O), fraction of solar radiation absorbed at the water surface (BETA), blue-green algae maximum growth rate (AG), blue-green algae maximum respiration rate (AR), blue-green algae maximum mortality rate (AM), blue-green algae settling rate (AS), blue-green algae light saturation intensity (ASAT), blue-green algae upper temperature for the rising rate function (AT2), labile dissolved organic matter decay rate (LDOMDK), labile particulate organic matter decay rate (LPOMDK), particulate organic matter settling rate (POMS), ammonia nitrification rate (NH4DK), release rate of ammonia from the sediment (NH4R), denitrification rate (NO3DK), release rate of phosphorus from the sediment (PO4R), and the three reaeration coefficients.

Water temperature was relatively insensitive to many model parameters, including WSC, EXH2O, and BETA; water temperature was instead more strongly controlled by the temperature of the inflows and overall meteorological conditions. The EXH2O parameter did affect dissolved oxygen, blue-green algae, ammonia, and orthophosphorus concentrations, showing that algal production was often light limited in this system. As expected, blue-green algae concentrations also were sensitive to manipulation of rates of growth, mortality, and settling, with less sensitivity to the respiration rate and temperature factors. An increase in blue-green algae growth rates produced increases in dissolved oxygen and ammonia and a decrease in orthophosphorus. The effect of changes in WSC and two of the reaeration coefficients on dissolved-oxygen concentrations demonstrates the importance of reaeration to the overall oxygen budget.

Model sensitivity, coupled with calibration and testing, aids in considering model uncertainty. Models are abstractions of reality and model equations necessarily involve simplified representations of complex natural phenomenon. The type of uncertainty associated with simplified representations can be termed "structural uncertainty" and can be minimized, but not eliminated, by choosing a model that has well-tested algorithms for the most important water-quality processes. Experimental work to better understand the dynamics of the most important processes also can help to refine algorithms and reduce such uncertainty. Important processes in the Upper Klamath River include algal and organic matter dynamics, SOD, and reaeration. Another source of model uncertainty is associated with input data. Uncertainty was minimized in this study by directly measuring as many of the needed inputs and calibration data as possible, with high-quality field and laboratory procedures, and using supported assumptions for the estimation of other unmeasured inputs. Finally, the selection of model parameters can be another source of model uncertainty. The PEST optimization software in this study was used to help choose a parameter set that minimized the error between model output and measured calibration data. These values also were compared to ensure parameters were within the range of values reported in published literature. In sum, it is important to realize that all models exhibit uncertainty that cannot be wholly eliminated. The approach for this project was to (1) minimize uncertainty through well-developed physical, chemical, and biological model representations; (2) provide careful field data-collection efforts; and (3) perform intensive model calibration and testing.

Table 7.   Results from sensitivity testing demonstrating the effect of changing model parameter values by 20 percent for the period June through October 2007 for the Klamath River upstream of Keno Dam, Oregon.

[See tables 4–5 for more details on each parameter]

Model parameter or input		Model output Whole reach, percent change June–October				
Parameter	Percent change	Water temperature	Dissolved oxygen	Blue-green algae	Ammonia	Ortho-phosphorus
WSC	-20	0	-6	1	-1	-1
	+20	-1	4	0	0	0
EXH2O	-20	0	4	5	3	-2
	+20	0	-3	-4	-2	1
BETA	-20	0	0	0	0	0
	+20	0	0	0	0	0
AG, blue-green	-20	0	-10	-16	-9	4
	+20	0	12	19	10	-5
AR, blue-green	-20	0	1	2	-1	-1
	+20	0	-1	-2	1	0
AM, blue-green	-20	0	4	19	7	-1
	+20	0	-3	-13	-5	1
AS, blue-green	-20	0	9	16	-3	-4
	+20	0	-7	-13	2	3
ASAT, blue-green	-20	0	0	0	0	0
	+20	0	0	0	0	0
AT2, blue-green	-20	0	2	-1	1	-1
	+20	0	-4	1	-3	2
LDOMDK	-20	0	2	0	-1	-1
	+20	0	-1	0	1	0
LPOMDK	-20	0	4	-1	-4	-1
	+20	0	-3	1	3	1
POMS	-20	0	3	-1	1	0
	+20	0	-3	1	-1	-1
NH4DK	-20	0	0	0	0	0
	+20	0	0	0	0	0
NH4R	-20	0	0	0	0	0
	+20	0	0	0	0	0
NO3DK	-20	0	0	0	0	0
	+20	0	0	0	0	0
PO4R	-20	0	0	0	0	0
	+20	0	7	0	0	0
Reaeration coefficient 1	-20	0	0	0	0	0
	+20	0	0	0	0	0
Reaeration coefficient 2	-20	0	-2	0	0	0
	+20	0	2	0	0	0
Reaeration coefficient 3	-20	0	-4	0	0	0
	+20	0	6	0	0	0

Model Application

The calibrated models were used to run a set of scenarios to examine the water-quality changes that might occur in this reach if inputs from the three point sources (Klamath Falls WWTP, South Suburban WWTP, Columbia Forest Products) and two tributaries (Lost River Diversion Channel, Klamath Straits Drain) were altered so that they were in compliance with the Klamath River TMDL (Oregon Department of Environmental Quality, 2010). The effect of altering the Link River input to meet Upper Klamath Lake TMDL in-lake phosphorus targets (Oregon Department of Environmental Quality, 2002) also was examined. The following four scenarios were run for calendar years 2006–09:

1a. Base case; Link River, point sources, and tributaries at existing conditions.

1b. Link River at existing conditions; point sources and tributaries in compliance with TMDL.

2a. Link River at TMDL targets; point sources and tributaries at existing conditions.

2b. Link River at TMDL targets; point sources and tributaries in compliance with TMDL.

Scenario Setup

Point Sources and Tributaries

Wasteload allocations from the Klamath River TMDL (table 8; Oregon Department of Environmental Quality, 2010) were used to adjust the model input files for Klamath Falls WWTP, South Suburban WWTP, Lost River Diversion Channel, and Klamath Straits Drain. ODEQ staff were consulted to determine how to most appropriately adjust the source loadings in light of the TMDL allocations. The first two source loadings are considered point sources, with allocations regulated as monthly medians. The last two source loadings are considered nonpoint sources in the TMDL, and those allocations were considered as yearly averages. Columbia Forest Products met all TMDL allocations under existing conditions, so this input was unaltered in the scenarios. In adjusting the input files, constituents that contributed to total phosphorus (orthophosphorus, dissolved organic matter, particulate organic matter, and the three algae groups) were first decreased to meet the phosphorus loading criteria, which required the largest reductions (table 9). The percent reduction required to meet the allocations were held constant through each year, thus maintaining seasonal patterns, but allowing variation between years. It was assumed that DOM would be more difficult to remove, so for the Lost River Diversion Channel and Klamath Straits Drain, which have wetlands and associated DOM from their tributaries, model DOM concentrations were not reduced as much as the other

Table 8. Load and wasteload allocations from the Klamath River, Oregon, total maximum daily load (TMDL) program.

[From Oregon Department of Environmental Quality (2010). **Abbreviations:** BOD5, 5-day biochemical oxygen demand; WWTP, wastewater treatment plant; lb/d, pound per day]

Source	TMDL allocation (lb/d)		
	BOD5	Total phosphorus	Total nitrogen
Point source			
Klamath Falls WWTP	488	9.6	618
South Suburban WWTP	308	6.0	390
Columbia Forest Products	40	2.1	10
Nonpoint source			
Lost River Diversion Channel[1]	2,998	42	546
Klamath Straits Drain	1,329	21	268

[1]For periods when flow is towards the Klamath River.

constituents. Their reduction value was usually 10 percent lower than reductions for other phosphorus-containing constituents.

Because organic matter and algae contain nitrogen as well as phosphorus, reductions to meet the phosphorus allocations also reduced inputs of nitrogen. If further reductions to meet nitrogen loading criteria were needed, then nitrate and ammonia concentrations were decreased. The TMDL BOD5 allocations (table 8) were met after phosphorus reductions and did not require additional reductions in input concentrations. The model that was used to set TMDL allocations (Tetra Tech, 2009) considered all organic matter in the inputs to be labile, resulting in a higher oxygen demand over short periods compared to the model described herein, which divides the input organic matter into labile and refractory groups based on more recent information on the nature of organic matter in this system.

Link River

In-lake phosphorus target criteria from the Upper Klamath Lake TMDL (Oregon Department of Environmental Quality, 2002) were used to produce a TMDL compliant Link River input. These target phosphorus concentrations were 30 ppb (µg/L) from March to May and 110 ppb from June to July. Although the TMDL describes the 110 ppb level as an annual lake target, it was considered herein as a summer target (Daniel Turner, Oregon Department of Environmental Quality, oral commun., December 10, 2010). Constituents that contributed to phosphorus in the Link River input (orthophosphorus, dissolved organic matter, particulate organic matter, the three algae groups) were decreased in concentration until the phosphorus concentration targets were met (table 9). Dissolved organic matter was reduced

Table 9. Percent reduction in Link River concentrations needed to meet Upper Klamath Lake, Oregon total maximum daily load (TMDL) in-lake total phosphorus concentration targets, and percent reduction in tributary concentrations needed to meet Klamath River TMDL total nitrogen and total phosphorus wasteload allocations.

[No reductions in Columbia Forest Products input concentrations were needed to meet TMDL allocations. **Abbreviations:** NH4, ammonia; NO3, nitrate; PO4, orthophosphorus; LPOM, labile particulate organic matter; RPOM, refractory particulate organic matter; LDOM, labile dissolved organic matter; RDOM, refractory dissolved organic matter; WWTP, wastewater treatment plant]

Year	Percent reduction for scenario model inputs			Year	Percent reduction for scenario model inputs		
	PO4, LPOM, RPOM, Algae (3 groups)	LDOM, RDOM	NH4, NO3		PO4, LPOM, RPOM, Algae (3 groups)	LDOM, RDOM	NH4, NO3
Link River				Lost River Diversion Channel			
2006	72	62	0	2006	94	84	91
2007	78	68	0	2007	75	65	42
2008	72	62	0	2008	79	69	47
2009	74	64	0	2009	59	49	0
Klamath Falls WWTP				Klamath Straits Drain			
2006	90	90	6	2006	98	90	92
2007	89	89	5	2007	93	83	84
2008	88	88	0	2008	94	84	89
2009	87	87	0	2009	93	83	83
South Suburban WWTP							
2006	95	95	0				
2007	93	93	0				
2008	93	93	20				
2009	93	93	0				

less than other constituents because Upper Klamath Lake has extensive wetlands in its drainage; those reduction values were 10 percent lower than for other phosphorus-containing constituents. The reduction factors were held constant throughout the year to retain a seasonal pattern that was similar to existing conditions. Because the seasonal pattern was retained, the 30 ppb criterion was more restrictive than the 110 ppb criterion.

Scenario Results

Dissolved Oxygen

Reducing concentrations of Link River inputs to reflect attainment of Upper Klamath Lake TMDL in-lake phosphorus targets (scenario 2a) was highly effective in increasing dissolved-oxygen concentrations in the Klamath River upstream of Keno Dam (table 10 summarizes all years; figure 21A shows 2007 as a representative year). This scenario resulted in average increases of 1.9 to 3.2 mg/L dissolved oxygen for the June through October period compared to the base case (scenario 1a, Link River, point sources, and

tributaries at existing conditions), but far more than that at certain locations when hypoxia occurred in the base case. In large part, this was caused by the large load reduction of blue-green algae and LPOM flowing into the reach from Link River and the resultant decrease in first-order SOD (fig. 21B–D). Blue-green algae concentrations decreased an average of 0.9–1.9 mg/L from June through October in scenario 2a compared to the base case (scenario 1a). Concentrations of other algae groups also decreased in the Link River inflow, but the algae were dominated by blue-greens in summer. Dead and decomposing blue-green algae and LPOM exerted a substantial oxygen demand in the water column and in the first-order sediment compartment after that material settles to the river bottom. Although the overall result was an increase in dissolved oxygen, an examination of spatial and temporal trends (fig. 21A) also showed a minor decrease in dissolved-oxygen concentration in late June and early July. That was a period when the blue-green algae were producing a substantial amount of oxygen through photosynthesis under existing conditions, so the simulated decrease occurred when oxygen was relatively abundant.

Table 10.   Annual and June through October volume average concentrations of dissolved oxygen, ammonia, orthophosphorus, and blue-green algae for Link River to Keno Dam, Oregon, as simulated by model scenarios 1 and 2.

[**Scenario:** KR, Klamath River; UKL, Upper Klamath Lake; TMDL, total maximum daily load. **Abbreviations:** Oct., October; mg/L, milligrams per liter]

Scenario	In-reach volume average (mg/L)							
	Dissolved oxygen		Ammonia		Orthophosphorus		Blue-green algae	
	Annual average	June–Oct. average	Annual average	June–Oct. average	Annual average	June–Oct. average	Annual average	June–Oct. average
2006								
1a Base case	8.08	5.22	0.420	0.478	0.076	0.095	1.30	2.79
1b KR TMDL tributaries	8.17	5.37	0.390	0.456	0.042	0.064	1.27	2.73
2a UKL TMDL target Link River	9.11	7.07	0.306	0.241	0.049	0.049	0.47	1.02
2b KR TMDL tributaries and UKL TMDL target Link River	9.17	7.17	0.274	0.213	0.015	0.019	0.41	0.89
2007								
1a Base case	7.58	5.04	0.557	0.594	0.073	0.110	1.25	2.91
1b KR TMDL tributaries	7.63	5.08	0.539	0.576	0.046	0.080	1.21	2.83
2a UKL TMDL target Link River	8.79	7.38	0.409	0.259	0.042	0.051	0.42	0.98
2b KR TMDL tributaries and UKL TMDL target Link River	8.82	7.38	0.388	0.236	0.014	0.021	0.37	0.85
2008								
1a Base case	6.79	3.83	0.641	0.456	0.071	0.088	0.95	2.18
1b KR TMDL tributaries	6.86	3.90	0.615	0.435	0.043	0.060	0.92	2.11
2a UKL TMDL target Link River	8.33	6.74	0.536	0.231	0.046	0.049	0.38	0.88
2b KR TMDL tributaries and UKL TMDL target Link River	8.38	6.79	0.507	0.204	0.018	0.021	0.33	0.76
2009								
1a Base case	6.98	3.36	0.528	0.463	0.059	0.076	0.66	1.49
1b KR TMDL tributaries	7.03	3.42	0.516	0.453	0.036	0.054	0.65	1.46
2a UKL TMDL target Link River	8.59	6.53	0.440	0.266	0.038	0.039	0.25	0.56
2b KR TMDL tributaries and UKL TMDL target Link River	8.64	6.58	0.426	0.251	0.015	0.017	0.22	0.50

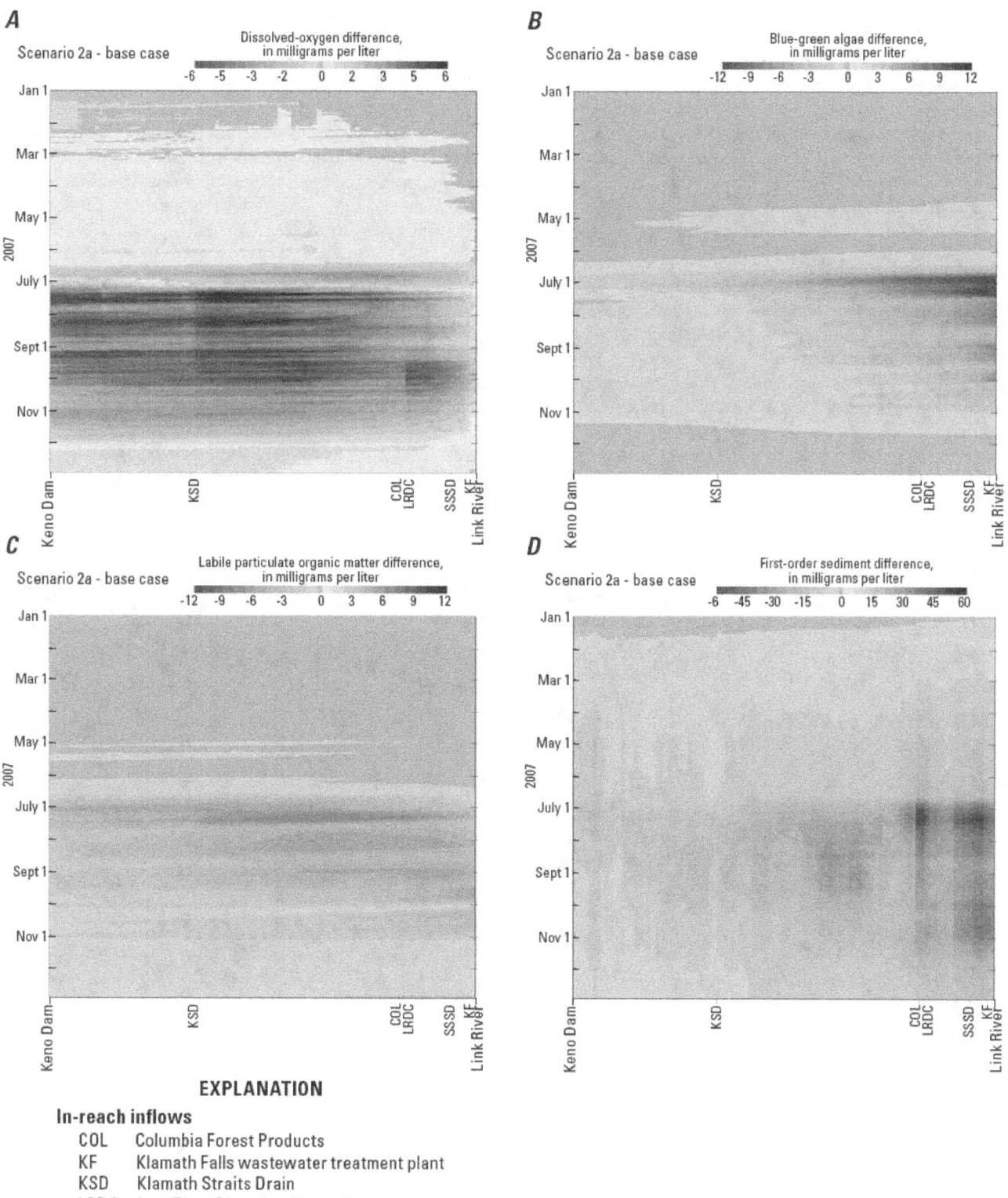

Figure 21. Temporal and spatial differences in modeled daily average concentrations of (A) dissolved oxygen, (B) blue-green algae, (C) labile particulate organic matter (LPOM) and (D) first-order sediment between the base scenario (1a) and the scenario with Link River meeting Upper Klamath Lake, Oregon, total maximum daily load (TMDL) in-lake phosphorus targets (2a) for 2007. Results are presented as the concentrations in scenario 2a minus concentrations in the base case (scenario 1a).

Reducing other tributary inputs to be compliant with the Klamath River TMDL (scenario 1b) also improved dissolved-oxygen concentrations in the reach, but to a lesser extent. This scenario increased dissolved oxygen for the June through October period by 0.0–0.2 mg/L, depending on the year. The combined effect of reducing concentrations from Link River and in-reach tributaries (scenario 2b) increased the reach-averaged dissolved-oxygen concentration by 2.0–3.2 mg/L for the June through October period.

Ammonia

Among the tested scenarios, reducing concentrations of Link River inputs to be compliant with Upper Klamath Lake TMDL requirements (scenario 2a) also was most effective in decreasing ammonia concentrations (table 10 summarizes all years, fig. 22A shows 2007 as a representative year). The June through October average decrease in ammonia concentration ranged from 0.20 to 0.34 mg/L depending on year for scenario 2a compared to the base case (scenario 1a). A

minor increase in nitrate concentrations resulted for the same time period (fig. 22B), due to a decrease in anoxic conditions that then allowed for some ammonia nitrification. Reductions of concentrations in the point and nonpoint sources (scenario 1b) had a more limited effect on ammonia concentrations, with average June to October decreases between 0.01 and 0.03 mg/L. Only minor decreases in nitrate concentrations occurred for that scenario. None of the scenarios had much of an effect on the high concentrations of ammonia that occurred during some years in winter.

Orthophosphorus

The scenarios that reduced inflow concentrations at Link River (2a and 2b) and the scenarios that reduced inflow concentrations for the in-reach tributaries (1b and 2b) produced substantial decreases in orthophosphorus concentrations through the system (table 10 summarizes all years, fig. 23 shows a representative year). Scenario 2a reduced orthophosphate concentrations between 0.04 and

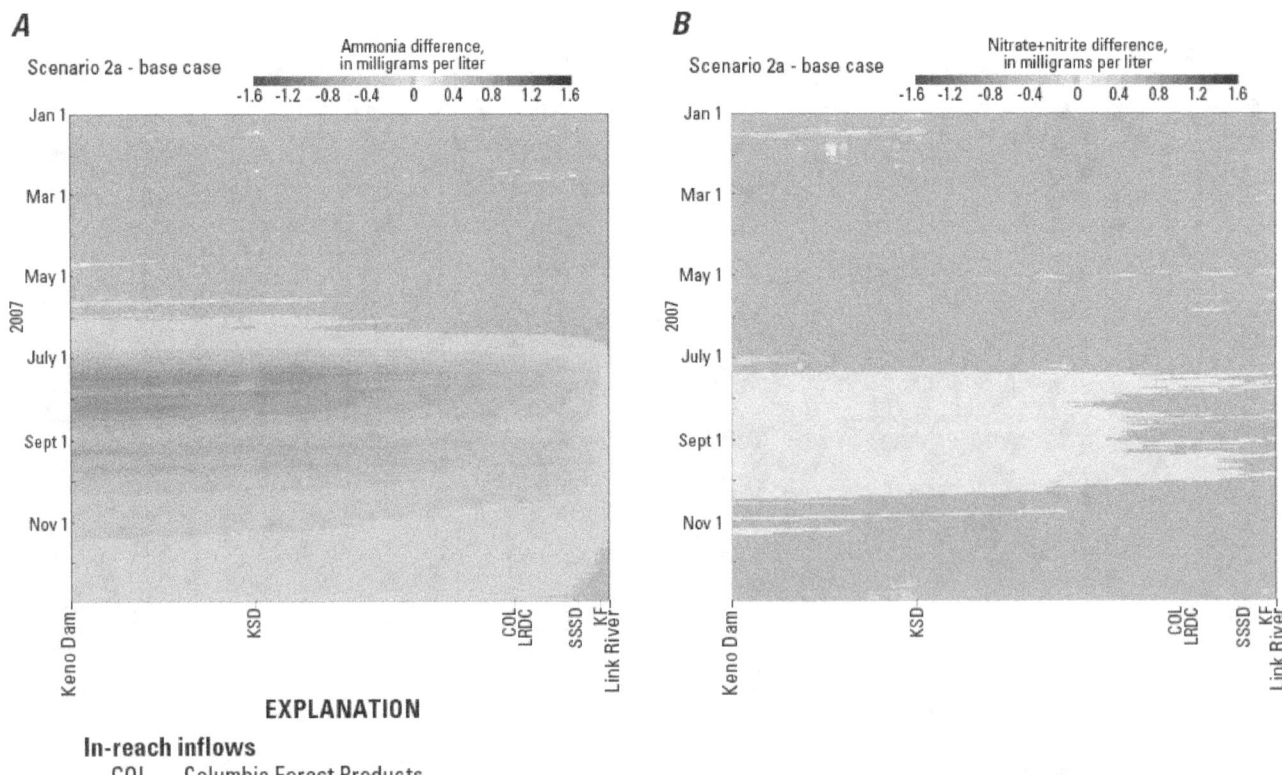

EXPLANATION

In-reach inflows
COL   Columbia Forest Products
KF   Klamath Falls wastewater treatment plant
KSD   Klamath Straits Drain
LRDC   Lost River Diversion Channel
SSSD   South Suburban wastewater treatment plant

Figure 22. Temporal and spatial differences in modeled daily average concentrations of (A) ammonia, and (B) nitrate plus nitrite between the base case scenario (1a) and the scenario with Link River meeting Upper Klamath Lake, Oregon, total maximum daily load (TMDL) in-lake phosphorus targets (2a) for 2007. Results are presented as concentrations in scenario 2a minus concentrations in the base case (scenario 1a).

0.06 mg/L for the June through October period compared to the base case, depending on year; scenario 1b reduced orthophosphate concentrations between 0.02 and 0.04 mg/L.

Scenarios Summary

This set of scenarios demonstrated that hypoxic and anoxic conditions in this reach of the Klamath River were largely a result of organic matter and algae imported from Upper Klamath Lake through Link River. If improvements in Upper Klamath Lake and Link River water quality are not possible over the short term, then the Link River to Keno reach of the Klamath River will continue to have high levels of seasonal water-quality impairment in response to large organic loads emanating from Upper Klamath Lake. Near-term management strategies focused on improving water quality within the Link–Keno reach may reduce the impacts of these loads and may result in water-quality improvements not only in the Link–Keno reach but also in the lower Klamath River

below Keno Dam. Although improvements in Upper Klamath Lake and Link River will have the largest influence on the presence and magnitude of anoxia in the Link–Keno reach, opportunities to decrease concentrations of nutrients, algae, particulate organic matter, and(or) algae in other tributary inputs could result in water-quality improvements as well.

Rather than retaining the current seasonal patterns of nutrients and organic matter, reduction factors could be adjusted on a daily or monthly basis for each tributary input for future model scenarios. Other management or restoration options, such as different methods of managing streamflow, treatment for point and nonpoint sources, treatment wetlands, and other prescriptions, could be examined. Future uses of the model could include connecting water-quality results to the water-quality requirements of fish, testing the potential effects of the removal of the four dams downstream of Keno Dam, assessing the effects of climate change and future water allocations, and examining downstream effects of potential changes to the Upper Klamath Lake TMDL.

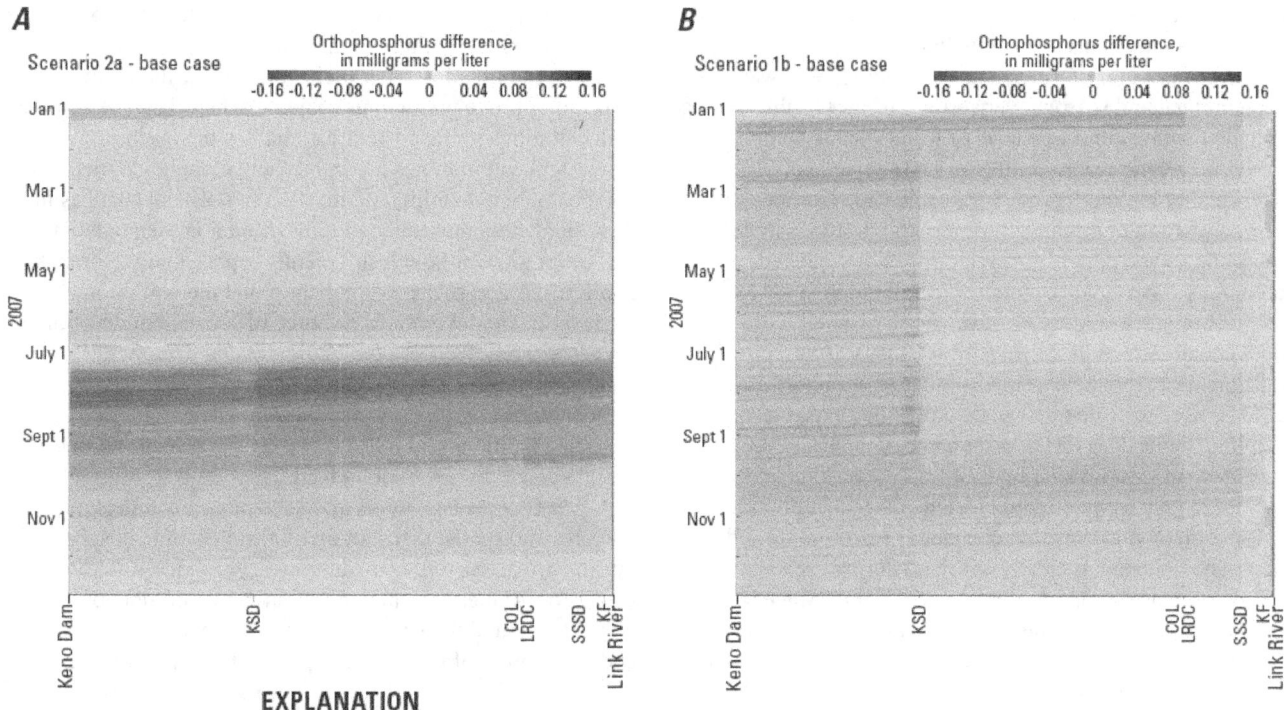

EXPLANATION

In-reach inflows

COL   Columbia Forest Products
KF   Klamath Falls wastewater treatment plant
KSD   Klamath Straits Drain
LRDC   Lost River Diversion Channel
SSSD   South Suburban wastewater treatment plant

Figure 23. Temporal and spatial differences in modeled daily average concentrations of orthophosphorus between the base case scenario (1a) and (A) the scenario with Link River meeting Upper Klamath Lake, Oregon, total maximum daily load (TMDL) in-lake phosphorus targets (2a), and (B) the scenario with in-reach point and nonpoint sources at Klamath River, Oregon, TMDL compliance (1b) for 2007. Results are presented as concentrations in scenario 2a or 1b minus concentrations in the base case (scenario 1a).

Implications for Future Monitoring and Management

Through the construction, calibration, and application of the models, a great deal was learned about the characteristics of the Klamath River system, the nature of the organic matter, the relative importance of algae, and the dynamic nature of oxygen demands. Interestingly, years with fewer input data available still produced good model results. For example, the error statistics for dissolved oxygen in 2006 and 2009 (years with fewer data) were comparable to 2007 and 2008 (years with rich datasets). The relatively good fit to the data in 2006 and 2009, however, could not have been achieved without the weekly field sampling and directed experimental work that occurred in 2007 and 2008, all of which was crucial to derive model rates and ratios and improve our understanding of spatial and temporal water-quality patterns in the Link–Keno reach. Future improvements to these and other models will rely on an increased understanding of the complex water-quality processes in this reach of the Klamath River. On the basis of what was learned in this modeling study, suggestions for future monitoring activities include:

Minimum level of water-quality sampling. Knowledge of interannual and seasonal variability in water quality can be expanded with a minimum level of water-quality sampling in the Link–Keno reach. Both grab samples and continuous water-quality monitors provide valuable and relevant data. A minimum program might include monthly to twice-monthly grab samples and data from a selected set of deployed continuous water-quality monitors. A minimum set of sample locations might include Link River, Miller Island, Keno, and Klamath Straits Drain. Ideally, telemetry would provide real-time data from at least one continuous monitor(s) that could be used to scale up or scale-down the frequency of sampling, depending on the occurrence of unusual events or critical periods of water quality. Useful constituents from these grab samples could include those that provide information on nutrients, organic matter partitioning, and algae. The field data typically collected during each site visit, including vertical profiles, algal density, turbidity, ice cover, and Secchi depth, could aid future analyses and modeling. The inclusion of samples in winter would be valuable because less historical information exists for that period. Measurements of inorganic sediment concentrations during periods of high flow could expand the equation relating turbidity and suspended sediment. The details of any water-quality monitoring plan need to be geared toward providing data to meet specific objectives. A monitoring plan to provide data for modeling purposes, for example, may require more frequent data collection at more sites than might be required otherwise.

Channel bottom sampling and sedimentation rates. The first-order sediment compartment in the model simulates the spatially variable buildup and decay of algal and organic material that settles to the river bottom. Sampling of material at the sediment–water interface in summer and winter could support documentation of this process in the field. Continued work on characterizing inflowing organic matter, sediment deposition rates, and additional studies on the potential for resuspension of sediments, composition of settled materials, decay rates under *in-situ* conditions, and other factors could greatly aid any future model refinements for sediment-related processes and shed light on dissolved-oxygen conditions. In addition to seasonal variation, sufficient spatial characterization is important when assessing sediment conditions.

Macrophyte survey and algae studies. High densities of rooted aquatic plants (macrophytes) occur only in certain areas of the Link–Keno reach due in part to the low penetration of light through the water column. However, some areas in the downstream half of the reach do have macrophyte populations. A series of macrophyte surveys, including density and species, would be highly beneficial to characterize the extent of these plants, assess their relative importance to the water quality of the river, and allow their inclusion into the model in an effort to improve predictions of near-surface dissolved-oxygen concentrations in the downstream half of the reach.

Algal communities are immensely complex in their composition, dynamics, and reaction to seasonal changes in flow, light, and nutrient conditions. Model implementation and calibration, although sufficiently representative for many applications and scenarios, indicates that the representation of algae in the model remains a source of substantial uncertainty. A special study focusing on longitudinal and temporal conditions in the Link–Keno reach could lend insight into these important processes, and their impact on dissolved-oxygen conditions and other water-quality effects, particularly in light of future restoration efforts.

Nutrient studies. Experiments focused on understanding the rates and spatial distribution of nitrogen and phosphorus cycling processes will contribute to better models. Potential studies could measure the rates of ammonia production, nitrification, and denitrification, and also examine the forms and availability of phosphorus species in the water column. The relationships of nutrients and primary production play a critical role in the water-quality conditions in this reach.

Link River. The model developed in this study does not include the 1-mi Link River reach. Although the Link River reach is short, questions have arisen about potentially important nutrient transformations there. Before developing a model of the 1-mi Link River reach, additional nutrient and organic matter sampling is needed at several sites—upstream of Link Dam, the outflow of Link Dam, and at the mouth of Link River—to document any such water-quality transformations. This piece of the model could become important if the need exists to connect the Link–Keno Klamath River model to an Upper Klamath Lake model.

Reaeration equations for Keno Dam. The model does not yet simulate the reaeration of dissolved oxygen at Keno Dam. For modeling, equations relating dissolved-oxygen concentrations and spill rates would need to be developed specific to Keno Dam. This addition would become important to connect the Link–Keno model to downstream models to simulate dissolved oxygen in the lower reaches of the Klamath River.

Acknowledgments

This project was made possible with primary funding from the Bureau of Reclamation and additional funds from U.S. Geological Survey (USGS). This was a project with many talented contributors including: Damion Ciotti, Matthew Kritzer, Scott Miller, James Ross, Gunter Schanzenbacher, Bill Stroud, and April Tower from the Bureau of Reclamation; William Ayers, Jon Baldwin, Amy Brooks, Kenna Butler, Amari Dolan-Caret, Elizabeth Jones, Julie Kirshtein, Laura Lambert, William Lehman, Mary Lindenberg, Jacqueline Olson, Dean Snyder, Steven Sobieszczyk, and Mary Voytek from the U.S. Geological Survey; Jim Sweet of Aquatic Analysts; Allan Vogel from ZP's Taxonomic Services; and Simon Poulson from the University of Nevada, Reno. Discussions with Jason Cameron, Rick Carlson, Jon Hicks, and Chuck Korson from the Bureau of Reclamation; Kurt Carpenter, Dennis Lynch, Joe Rinella, and Tammy Wood from the U.S. Geological Survey; Daniel Turner from Oregon Department of Environmental Quality; and John Doherty from Watermark Numerical Consulting were helpful at various stages of the project. Comments by two peer reviewers improved the report.

References Cited

American Society of Civil Engineers, 2000, Hydraulic modeling: Concepts and practice: ASCE Manuals and Reports in Engineering, 390 p.

Auer, B., Elzer, U., and Arndt, H., 2004, Comparison of pelagic food webs in lakes along a trophic gradient and with seasonal aspects: Influence of resource and predation: Journal of Plankton Research, v. 26, no. 26, p. 679–709.

Boers, P., Van Ballegooijen, L.V., and Uunk, J., 1991, Changes in phosphorus cycling in a shallow lake due to food web manipulations: Freshwater Biology, v. 25, p. 9–20.

Bradbury, J.P., Colman, S.M., and Reynolds, R.L., 2004, The history of recent limnological changes and human impact on Upper Klamath Lake, Oregon: Journal of Paleolimnology, v. 31, p. 151–165.

Canuel, E.A., and Martens, C.S., 1996, Reactivity of recently deposited organic matter: Degradation of lipid compounds near the sediment-water interface: Geochimica et Cosmochimia Acta, v. 60, p. 1793–1806.

CH2M-Hill and Wells, S.A., 1995, Water quality model of the Klamath River between Link River and Keno Dam— Draft report for the Oregon Department of Environmental Quality: Portland, Oregon [variously paged], accessed May 20, 2011, at http://or.water.usgs.gov/proj/keno_reach/download/CH2MHill_Wells_1995_draft.pdf.

Cole, T.M., and Wells, S.A., 2008, CE-QUAL-W2—A two-dimensional, laterally averaged, hydrodynamic and water-quality model, version 3.6: U.S. Army Corps of Engineers, Instruction Report EL-08-1, [variously paged].

De Nobel, W.T., Matthijs, H.C.P., von Elert, E., and Mur, L.R., 1998, Comparison of the light-limited growth of the nitrogen-fixing cyanobacteria *Anabaena* and *Aphanizomenon*: New Phytologist, v. 138, no. 4, p. 579-587.

Deas, M.L., and Vaughn, J., 2006, Characterization of organic matter fate and transport in the Klamath River below Link Dam to assess treatment/reduction potential—Prepared for the Bureau of Reclamation, Klamath Area Office, September 30, 2006: Davis, California, Watercourse Engineering, Inc., 167 p.

Deas, M.L., and Vaughn, J., 2011, Keno Reservoir particulate study 2008—Technical memorandum, Prepared for the Bureau of Reclamation, Klamath Basin Area Office, April 2011: Davis, California, Watercourse Engineering, Inc., 38 p.

Deas, M.L., Vaughn, J., and Tanaka, S., 2009, Exploratory assessment of stable isotopes in the Klamath River— Technical memorandum, Prepared for PacifiCorp: Davis, California, Watercourse Engineering, Inc., 10 p.

Doherty, J., 2010, PEST: Model—Independent parameter estimation, user manual (5th ed.): Brisbane, Australia, Watermark Numerical Computing, 336 p.

Doyle, M.C., and Lynch, D.D., 2005, Sediment oxygen demand in Lake Ewauna and the Klamath River, Oregon, June 2003: U.S. Geological Survey Scientific Investigations Report 2005–5228, 14 p. (Also available at http://pubs.usgs.gov/sir/2005/5228/pdf/sir2005-5228.pdf.)

Dunsmoor, L.K., and Huntington, C.W., 2006, Suitability of environmental conditions within Upper Klamath Lake and the migratory corridor downstream for use by anadromous salmonids—Technical memorandum to the Klamath Tribes, October 2006: Chiloquin, Oregon, 80 p.

Eilers, J.M., and Gubala, C.P., 2003, Bathymetry and sediment classification of the Klamath hydropower project impoundments—Prepared for PacifiCorp: Bend, Oregon, J.C. Headwaters, Inc., 39 p., accessed on May 17, 2011, at http://www.pacificorp.com/content/dam/pacificorp/doc/Energy_Sources/Hydro/Hydro_Licensing/Klamath_River/Bathymetry_and_Sediment_Classification_Report_Final_April_2003.pdf.

Eilers, J.M., Kann, J., Cornett, J., Moser, K., and St. Amand, A., 2003, Paleolimnological evidence of change in a shallow hypereutrophic lake—Upper Klamath Lake, Oregon, USA: Hydrobiologia, v. 520, p. 7–18.

Eilers, J.M., and Raymond, R., 2003, Sediment oxygen demand and nutrient release from sites in the Klamath Hydropower Project—A report to PacifiCorp: Bend, Oregon, J.C. Headwaters, Inc., 63 p.

Eilers, J.M., and Raymond, R., 2005, Sediment oxygen demand in selected sites of the Lost River and Klamath River—Prepared for Tetra Tech: Bend, Oregon, MaxDepth Aquatics, Inc., 17 p., accessed on May 17, 2011, at http://www.deq.state.or.us/wq/tmdls/docs/klamathbasin/uklost/KlamathLostAppendixE.pdf.

Gannett, M.W., Lite, K.E. Jr., La Marche, J.L., Fisher, B.J., and Polette, D.J., 2007, Ground-water hydrology of the upper Klamath Basin, Oregon and California: U.S. Geological Survey Scientific Investigations Report 2007–5050, 84 p. (Also available at http://pubs.usgs.gov/sir/2007/5050/.)

Hem, J.D., 1985, Study and interpretation of the chemical characteristics of natural water: U.S. Geological Survey Water-Supply Paper 2254, 263 p. (Also available at http://pubs.usgs.gov/wsp/wsp2254/.)

Hendrickson, J., Trahan, N., Gordon, E., and Ouyang, Y., 2007, Estimating relevance of organic carbon, nitrogen, and phosphorus loads to a Blackwater River estuary: Journal of the American Water Resources Association, v. 43, no. 1, p. 264–279.

Idso, S.B., and Gilbert, R.G., 1974, On the universality of the Poole and Atkins Secchi disk-light extinction equation: Journal of Applied Ecology, v. 11, p. 399–401.

Ji, Z-G., and Jin, K-R., 2006, Gyres and seiches in a large and shallow lake: Journal of Great Lakes Research, v. 32, p. 764–775.

Latja, R., and Salonen, K., 1978, Carbon analysis for the determination of individual biomass of planktonic animals: Verhandlungen der Internationalen Vereinigung für Limnologie, v. 20, p. 2556–2560.

Marine, K.R., and Lappe, P., 2009, Link River Dam surface spill experiment: Evaluation of differential juvenile sucker downstream passage rates—Final Report prepared for Bureau of Reclamation, March 2009, 68 p.

Mrazik, S., 2007, Oregon water quality index summary water years 1997-2006: Oregon Department of Environmental Quality, DEQ07-LAB-007-TR, 13 p.

Nagata, T., 1986, Carbon and nitrogen content of natural planktonic bacteria: Applied Environmental Microbiology, v. 52, p. 28–32.

National Oceanic and Atmospheric Administration, 1998, Automated Surface Observing System (ASOS) user's guide: accessed June 2, 2011, at http://www.nws.noaa.gov/asos/aum-toc.pdf.

Oregon Department of Environmental Quality, 2002, Upper Klamath Lake drainage total maximum daily load (TMDL) and water quality management plan (WQMP), May 2002: Portland, Oregon, 188 p., accessed June 2, 2011, at http://www.deq.state.or.us/wq/tmdls/docs/klamathbasin/ukldrainage/tmdlwqmp.pdf.

Oregon Department of Environmental Quality, 2007, Oregon's 2004/2006 integrated report, accessed November 16, 2007, at http://www.deq.state.or.us/wq/assessment/rpt0406.htm.

Oregon Department of Environmental Quality, 2009, Laboratory analytical storage and retrieval (LASAR) database, accessed May 16, 2011, at http://www.deq.state.or.us/lab/lasar.htm.

Oregon Department of Environmental Quality, 2010, Upper Klamath and Lost River subbasins total maximum daily load (TMDL) and water quality management plan (WQMP), December 2010, accessed May 17, 2011, at http://www.deq.state.or.us/WQ/TMDLs/klamath.htm.

Oregon Department of Fish and Wildlife, 2008, A plan for the reintroduction of anadromous fish in the Upper Klamath Basin (Draft), March 2008: Salem, Oregon, 53 p., accessed on May 17, 2011, at http://www.dfw.state.or.us/agency/commission/minutes/08/07_july/Exhibit%20B_Attachment%204.pdf.

PacifiCorp, 2002, Explanation of facilities and operational issues associated with PacifiCorp's Klamath hydroelectric project, FERC Project No. 2082 (Draft): Portland, Oregon, PacifiCorp, 44 p., accessed May 31, 2011, at http://www.pacificorp.com/content/dam/pacificorp/doc/Energy_Sources/Hydro/Hydro_Licensing/Klamath_River/Klamath_Project_Facilities_and_Operations_Report.pdf.

PacifiCorp, 2005, Klamath River water quality model implementation, calibration, and validation, Klamath Hydroelectric Project (FERC Project No. 2082)—Response to November 10, 2005, FERC AIR GN–2: Portland, Oregon, PacifiCorp, FERC filing: December 16.

Padisak, J., 2004, Phytoplankton, in O'Sullivan, P.E., and Reynolds, C.S., eds., The lakes handbook, volume 1, limnology and limnetic ecology: Oxford, Blackwell Science, p. 251–308.

Poole, H.H., and Atkins, W.R.G., 1929, Photo-electric measurements of submarine illumination throughout the year: Journal of the Marine Biological Association, v. 16, p. 297-324.

Poulson, S.R., and Sullivan, A.B., 2010, Assessment of diel chemical and isotopic techniques to investigate biogeochemical cycles in the Upper Klamath River, Oregon, USA: Chemical Geology, v. 269, no. 1–2, p. 3–11.

Raymond, R., and Eilers, J.M., 2004, Sediment oxygen demand and nutrient release from sites in Lake Ewauna and Keno Reservoir: MaxDepth Aquatics, Inc., 18 p.

Reynolds, C.S., 2006, The Ecology of Phytoplankton: Cambridge University Press, 550 p.

Risley, J.C., Hess, G.W., and Fisher, B.J., 2006, An assessment of flow data from Klamath River sites between Link River Dam and Keno Dam, south-central Oregon: U.S. Geological Survey Scientific Investigations Report 2006–5212, 30 p. (Also available at http://pubs.usgs.gov/sir/2006/5212/.)

Rocha, O., and Duncan, A., 1985, The relationship between cell carbon and cell volume in freshwater algal species used in zooplanktonic studies: Journal of Plankton Research, v. 7, no. 3, p. 279–294.

Rounds, S.A., and Sullivan, A.B., 2009, Review of Klamath River total maximum daily load models from Link River Dam to Keno Dam, Oregon: U.S. Geological Survey Administrative Report, 37 p., accessed November 18, 2010, at http://or.water.usgs.gov/proj/keno_reach/download/klamath_river_model_review_final.pdf.

Rounds, S.A., and Sullivan, A.B., 2010, Review of revised Klamath River total maximum daily load models from Link River Dam to Keno Dam, Oregon: U.S. Geological Survey Administrative Report, 32 p. (Also available at http://or.water.usgs.gov/proj/keno_reach/download/klamath_model_rereview_final.pdf.)

Sammel, E.A., 1980, Hydrogeologic appraisal of the Klamath Falls geothermal area, Oregon: U.S. Geological Survey Professional Paper 1044–G, 45 p.

Sullivan, A.B., Deas, M.L., Asbill, J., Kirshtein, J.D., Butler, K., Wellman, R.W., Stewart, M.A., and Vaughn, J., 2008, Klamath River water quality and acoustic Doppler current profiler data from Link River Dam to Keno Dam, 2007: U.S. Geological Survey Open-File Report 2008–1185, 25 p. (Also available at http://pubs.usgs.gov/of/2008/1185/.)

Sullivan, A.B., Deas, M.L., Asbill, J., Kirshtein, J.D., Butler, K., and Vaughn, J., 2009, Klamath River water quality data from Link River to Keno Dam, Oregon, 2008: U.S. Geological Survey Open-File Report 2009–1105, 25 p. (Also available at http://pubs.usgs.gov/of/2009/1105/.)

Sullivan, A.B., Snyder, D.M., and Rounds, S.A., 2010, Controls on biochemical oxygen demand in the Upper Klamath River, Oregon: Chemical Geology, v. 269, no. 1–2, p. 12–21.

Terwilliger, M.R., Simon, D.C., and Markle, D.F., 2004, Larval and juvenile ecology of Upper Klamath Lake suckers: 1998–2003: final report to Bureau of Reclamation, 217 p.

Tetra Tech, 2009, Klamath River model for TMDL development, Prepared for U.S. Environmental Agency Region 9 and 10, Oregon Department of Environmental Quality, and North Coast Regional Water Quality Control Board, December 2009: 196 p., accessed May 20, 2011, at http://www.deq.state.or.us/wq/tmdls/docs/klamathbasin/uklost/KlamathLostAppendixC.pdf.

U.S. Environmental Protection Agency, 1997, Analysis of the affect of ASOS-derived meteorological data on refined modeling: Office of Air Quality, EPA-454/R-97-014, November, 1997 [variously paged].

U.S. Environmental Protection Agency, 2003, Standard operating procedures for zooplankton analysis: LG403, 16 p.

Vaughn, J., and Deas, M., 2006, Keno Reservoir vertical, lateral and longitudinal data—Technical memorandum prepared for J. Cameron, Bureau of Reclamation, Klamath Basin Area Office, December 21, 2006: Davis, California, Watercourse Engineering, Inc., 83 p.

Vuorio, K., Meili, M., and Jouko, Sarvala, 2006, Taxon-specific variation in the stable isotopic signatures (δ^{13}C and δ^{15}N) of lake phytoplankton: Freshwater Biology, v. 51, p. 807–822.

Wagner, R.J., Boulger, R.W., Jr., Oblinger, C.J., and Smith, B.A., 2006, Guidelines and standard procedures for continuous water-quality monitors—Station operation, record computation, and data reporting: U.S. Geological Survey Techniques and Methods 1–D3, 51 p. plus 8 attachments. (Also available at http://pubs.usgs.gov/tm/2006/tm1D3/.)

Watercourse Engineering, Inc., 2003, Klamath River modeling framework to support the PacifiCorp Federal Energy Regulatory Commission Hydropower Relicensing Application, November 14, 2003: 291 p., accessed May 31, 2011, at http://www.pacificorp.com/content/dam/pacificorp/doc/Energy_Sources/Hydro/Hydro_Licensing/Klamath_River/WR_Appendix_4A_Klamath_River_Modeling_Framework.pdf.

Williams, D.T., Drummond, G.R., Ford, D.E., and Robey, D.L., 1980, Determination of light extinction coefficients in lakes and reservoirs, *in* Stefan, H.G. (ed.), Proceedings of the Symposium on Surface Water Impoundments, June 2–5, Minneapolis, Minnesota: New York, American Society of Civil Engineers, p. 1329–1335.

Wood, T.M., 2001, Sediment oxygen demand in Upper Klamath and Agency Lakes, Oregon, 1999: U.S. Geological Survey Water-Resources Investigations Report 01–4080, 13 p., http://or.water.usgs.gov/pubs/WRIR01-4080/.

Wood, T.M., Cheng, R.T., Gartner, J.W., Hoilman, G.R., Lindenberg, M.K., and Wellman, R.E., 2008, Modeling hydrodynamics and heat transport in Upper Klamath Lake, Oregon, and implications for water quality: U.S. Geological Survey Scientific Investigations Report 2008–5076, 48 p. (Also available at http://pubs.usgs.gov/sir/2008/5076/.)

Supplementary Materials

Visualizations of model-simulated results were produced from the existing conditions model runs for selected parameters and time periods. These visualizations help to convey an understanding of daily and seasonal variations as well as the relative importance of upstream and tributary inputs, vertical mixing, transport, and in-stream reactions. Animations using model-simulated results and the calibrated models can be accessed at the project web site, http://or.water.usgs.gov/proj/keno_reach/models.html.